Demystifying Chipmaking

Demystifying Chipmaking

*by Richard F. Yanda, Michael Heynes
and Anne K. Miller*

ELSEVIER

AMSTERDAM • BOSTON • HEIDELBERG • LONDON
NEW YORK • OXFORD • PARIS • SAN DIEGO
SAN FRANCISCO • SINGAPORE • SYDNEY • TOKYO
Newnes is an imprint of Elsevier

Newnes

Newnes is an imprint of Elsevier
30 Corporate Drive, Suite 400, Burlington, MA 01803, USA
Linacre House, Jordan Hill, Oxford OX2 8DP, UK

 Recognizing the importance of preserving what has been written,
Elsevier prints its books on acid-free paper whenever possible.

Library of Congress Cataloging-in-Publication Data

(Application submitted.)

British Library Cataloguing-in-Publication Data
A catalogue record for this book is available from the British Library.

ISBN: 0-7506-7760-0

For information on all Newnes publications
visit our Web site at www.books.elsevier.com

Contents

Contents

Foreword

Welcome! This is the story of how the vast majority of integrated circuits are made—call them "computer chips," if you like, although a host of other electronic components are made in much the same way. Everyone who is curious about how such tiny, yet powerful, devices are made will enjoy the story. This book will also serve as a valuable update to those who may be unaware of how drastically the state-of-the-art processes have changed in recent years.

The most glamorous (and expensive) part of the chipmaking process is the focus of this book. A chip will be built: each step is described, in order, until the whole chip is completed, from the design phase, to growing the silicon ingots, to the final testing of the packaged part.

The building of a chip is a fascinating process that will certainly impress and astonish everyone who is new to the story or who is out of touch with current procedures. Read on!

Acknowledgments

The authors are very grateful to these wonderful folks who helped make this project a success: Katy Yanda and Robert Lain of T-Ram, Rod Gross at Studio 144, David Craven of Lam Research Corporation, Patty Cimlov-Zahares, "Another Artist," and Karen Moore, and the analysts at VLSI Research.

About the Authors

Richard F. Yanda

Rick Yanda is a senior industry instructor with extensive experience in developing and delivering semiconductor-related courses. He served as manager as the Process Training Department at LAM Research where he developed the department curriculum and served as senior technical instructor. As founder of Semiconductor Technology Training Corp., he developed specialized industry training courses. Earlier in his career, Mr. Yanda was senior process engineer at Advanced Micro Devices, Cypress Semiconductor and INMOS Corporation. He also held the position of senior process engineer at Lam Research and Matrix Integrated Systems. Mr. Yanda holds a B.A. in Physics and has completed numerous post-graduate technical courses.

Michael S.R. Heynes, Ph. D

Dr. Heynes is senior engineer/scientist with broad semiconductor industry experience primarily in wafer fabrication. Since 1994, Dr. Heynes has been developing and delivering semiconductor-related courses. He was a senior technical instructor at LAM Research. He has worked extensively with Semiconductor Services as an instructor and course developer. He also in writes scripts for training videos in the area of semiconductor manufacturing as well as reviews and advises on content of technical articles. His industry experience includes management of engineering groups and small wafer fabrication operations. Dr. Heynes holds a Ph.D. in Physical Chemistry.

Anne K. Miller

Anne Miller has been managing Semiconductor Services since 1989. She is responsible for training product development and also provides technical marketing services to semiconductor-related firms. Ms. Miller had been the product marketing manager of the Wafer Inspection Division of KLA Instruments Corporation. Before that she was the product manager for the Semiconductor Equipment Division of Cambridge Instruments. As one of the pioneers in the development of pellicles, Ms. Miller founded Micropel, a subsidiary of EKC Technology. Other experience was gained as a process engineer at Intel Corporation, and product development manager at JASCO Chemical Corporation.

Before entering the semiconductor industry, Ms. Miller was a lecturer in the Chemistry Department at San Jose State University. She has an executive MBA from Stanford University, holds a junior college teaching credential in industrial technology, seventy post-graduate units in Business and Chemistry, and a B.S. degree in Chemistry. Ms. Miller has been a member of "Who's Who in the West" for many years.

What's on the CD-ROM?

The CD contains live footage from leading chip manufacturing facilities. It is the preview for "Making the Microchip—At the Limits III" video training course produced by Semiconductor Services in 2004. The video reflects current integrated manufacturing technology and includes topics such as shallow trench isolation, chemical mechanical planarization, low and high-k dielectrics and the dual damascene copper deposition process. Many of the images used in *Demystifying Chipmaking* are also in the video. It is possible for readers to rent or purchase the video training course from Semiconductor Services at www.semiconductorservices.com or view it at www.SemiZone.com.

CHAPTER

1

IC Fabrication Overview

Introduction

Chapter 1 contains a brief synopsis of the detailed discussion to come. It serves as an orientation and reference to give the reader a sense of the big picture and how much goes into the making of a "chip."

Figure 1-1: Example of an Integrated Circuit (Microprocessor)

1.1 Integrated Circuits

An integrated circuit (IC) is a miniaturized electronic device made by combining discrete electronic components into a single device, all manufactured as a whole. The IC may contain millions of transistors, diodes, resistors, capacitors, and other components integrated into a single chip.

Today, the term "chip" is often the favorite name given to ICs and will be used extensively throughout this discussion.

> The IC was originally called the *monolithic* integrated circuit. "Monolithic" means "single-stone," referring to the fact that the entire circuit is built on—and in—a single piece of "stone," the semiconducting material. In 1958, Robert Noyce (one of the founders of Intel), then at Fairchild Semiconductor and Jack Kilby of Texas Instruments were jointly awarded the patent for the IC, even though they did the work separately and at different companies. Noyce built his device on silicon and Kilby built his on germanium.
>
> The important difference in the two approaches to making the first IC was in the way they were wired together. Kilby used thermo-compression wire bonding to connect the components; Noyce used a patterning process and formed the wires by etching a thin aluminum film on the surface of the wafer. Noyce's technique became the standard method.

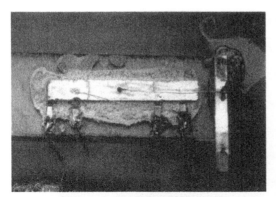

Figure 1-2a: Kilby's First IC

Figure 1-2b: Commercial Version of TI IC

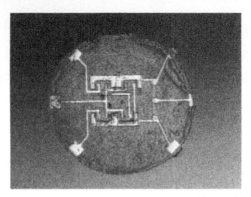

**Figure 1-2c: First IC Using Thin Film Wiring
(Noyce/Fairchild Semiconductor)**

Silicon (Si), the second most abundant element on earth, is the most familiar semiconductor material as well as the most widely used. Silicon technology is the subject of this book. The vast majority of all the ICs produced in the world today are made using silicon.

Germanium (Ge) is another familiar semiconductor. The first transistor was made of germanium and it was the principle material used for the first decade of the industry. It is still is of interest today.

Silicon and germanium are elements, but many semiconducting materials are chemical compounds formed between two or more chemical elements. Not surprisingly, they are called *compound semiconductors*. Gallium arsenide (GaAs), gallium nitride (GaN) and indium phosphide (InP) are examples of compound semiconductors. Although they are not as well recognized, they play a large role in several applications such as high frequency devices used in cell phones and in optoelectronics, most notably in fiber optics applications (telecommunications applications using glass fibers instead of copper wires). Compound semiconductors are also found in applications such as light emitting diodes (LEDs), which are used for traffic lights; they save energy and reduce maintenance costs, as well as producing more intense light.

Semiconductor materials make the magic happen for ICs. Pure semiconductors are not very good conductors of electricity. However, they have a quite handy property that allows them to become relatively good conductors if special impurities, called *dopants*, are added. Doped semiconductors possess either an excess of mobile negative charges (n-type), or an excess of (apparently) mobile positive charges (p-type) depending on the type of dopant chemical that is added. The mobile charges in n-type material are electrons. The mobile charge in a p-type conductor is called a *hole* and it is treated as though it were an actual positive charge carrier. The physics of semiconductor devices is very interesting, and the reader will find some good suggestions for further reading in the Bibliography.

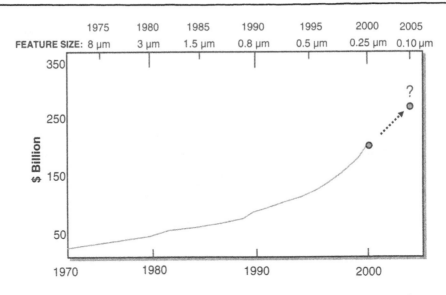

Figure 1-3: Semiconductor Industry Revenue with Chip Feature Sizes

Almost all of the devices made from semiconductors incorporate adjacent n-type and p-type regions in a variety of ways. The importance of putting these two oppositely doped regions right next to each other is discussed extensively throughout the book. In short, diodes, transistors of various types and other electrical components are made in this way.

1.2 The Semiconductor Industry

The semiconductor industry began in 1947 at Bell Laboratories in Murray Hill, NJ with the invention of the first transistor. At the time of this writing, the industry has grown to global proportions with annual chip sales exceeding $200 billion. The segment of the industry that makes processing equipment used to manufacture ICs is about $50 billion in annual sales.

Products include ICs, diodes, rectifiers and transistors made from semiconductor materials. ICs are now found in toasters, lamps, automobiles and most every electrical device. Of course, computers, office equipment, communications, defense, aerospace, industrial control and consumer electronics industries are dependent upon the semiconductor industry. Light emitting diodes (LEDs), lasers, photovoltaic (solar) cells, measurement and sensing devices are also part of the semiconductor industry.

Support Technologies

The focus of this book is the story of how ICs are produced using the principal technology employed today, CMOS. However, the complexity of this technology requires the support of other equally sophisticated technologies. The story would not be complete without including a discussion of these fascinating topics.

2.1 Crystal Growth and Wafer Preparation

Single crystal silicon is the material most commonly used in the fabrication of integrated circuits. The silicon is purified from a surprisingly familiar starting material: sand. Many sand deposits are primarily silicon dioxide. The preferred types of sand contain very low levels of impurities. This starting material is further purified and chunks of very pure polycrystalline silicon are extracted.

Next, the silicon is melted in a large quartz glass crucible. The furnace must reach a temperature of at least 1414°C (white heat), the melting point of silicon. A seed crystal in the form of a short rod about the size of a pencil with the desired atomic arrangement is lowered to touch the top of the melt. Silicon atoms attach to the seed crystal, slowly freeze and precisely replicate the atomic arrangement in the seed. The resulting ingot is gradually pulled up and out of the liquid melt. The ingot will often be 300 mm in diameter and one or two meters long.

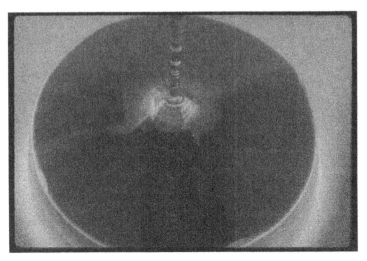

Figure 1-4: Silicon Ingot Being Pulled from Melt

The silicon ingot is machined on special tools that produce a more precisely shaped cylinder of the proper diameter. Then the ingot is cut into thin wafers using a diamond saw. The wafers are polished to a mirror finish using methods that minimize the damage done to the crystal structure at the surface. Most semiconductor devices use only a very shallow layer of the silicon at the wafer surface; maintaining good crystal properties is important to their performance.

2.2 Contamination Control

Integrated circuits must be manufactured in a carefully controlled environment. The devices are so small that even a bacterium contaminating the surface of the chip would likely cause it to fail. Dust particles much smaller than the eye can see will produce "killer" defects. Particles can result in defects in the tiny patterns formed in building the device structures. Particles could also be a source of trace chemical contamination that may damage the product.

ICs are made in cleanrooms, often called the *fabrication area* or *fab* for short. Contamination levels are classified in terms of the number of particles found in a cubic foot of air that are 0.5 μm or more in diameter. The μm stands for micrometer, or micron for short, a unit of measure that is one one-thousandth of a millimeter in length. The Class 1 cleanroom, common to the industry today, will only have one of those particles in each cubic foot of air. Class 10 and Class 100 cleanrooms are also useful for less demanding technologies.

Cleanroom Class Number	# of Particles > 0.5 $\mu m/ft^3$
1000	1000
100	100
10	10
1	1

Figure 1-5: Table of Cleanroom Classes

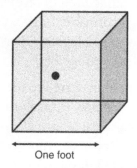

One foot

In a class-one cleanroom:
No more than one particle
0.5 micron or larger per cubic foot.

The cleanliness is maintained by the use of high efficiency particle attenuation (HEPA) filters. Ultra low particle attenuation (ULPA) filters are also used. These filters are positioned in the ceiling of the cleanroom or in the ceilings of special environmental chambers that must be kept contamination free. The air moves downward from the filters, producing a laminar flow pattern that will carry any stray particles downward through the return vents in or near the floor.

The cleanroom suit or "bunny suit" is critical to keeping the environment clean. It is a coverall made of nonparticulating material. The people working in the cleanroom are a major source of particles, so a complete coverall must be worn. Even makeup is forbidden because it produces a continuous cloud of particles emitted by the person wearing it.

The wafers upon which the chips are made are kept in closed containers during movement between manufacturing stations. The wafers are held in slotted cassettes, called boats, which keep them from touching each other. Automated robotic handlers are popular to move wafers through the fab. Each tool and inspection station is equipped with vacuum transfer arms or wands for handling individual wafers. The front of the wafer must never be touched except for processing!

2.3 Circuit Design and Mask Making

Chip design begins with a schematic diagram of the desired circuit. The schematic is translated into a picture or "layout" of the actual chip, drawn on a computer-aided design (CAD) tool. The chip is manufactured in layers; the overall layout is partitioned appropriately by the design software. Each layer is a pattern that must be reproduced on the wafer. These patterned layers, when stacked up, become the electronic components of the chip, all wired together.

The individual layer patterns are transferred to the chip using a photolithographic technique. A template, called a *photomask,* is made for each layer. The photomask is a picture or pattern etched in chrome on a quartz glass plate that allows light through in some areas and blocks the light elsewhere. Another form of the patterned plate is called a *reticle.* When light is projected through the photomask or reticle and onto a photosensitive material called *photoresist (resist)* coating the wafer, the pattern from the reticle will appear in the photosensitive material after treatment with a developer solution (see Section 3.2.3, Exposure, and Section 3.2.4, Develop). In effect, the photoresist-coated wafer acts like the film in a camera.

The reticles and photomasks are themselves made using a similar process. The mask-making company receives a computer file containing the design for each photomask or reticle. A blank, chrome-coated quartz plate is covered with photoresist. Then the plate is processed in a computer-controlled electron beam (e-beam) tool or a laser pattern generator that reproduces the desired pattern in the photoresist. Developer solution removes the unwanted resist. The remaining resist serves as the template (resist mask) for a wet etching process to remove the exposed chrome. When the pattern is reproduced in the chrome, the resist mask can be stripped off and the chrome mask (or reticle) is complete.

State-of-the-art chips require twenty to thirty such masks.

Photomasks are sometimes called *masks* for short so it can become confusing. Chapters 2 and 3 will clarify the terminology and elaborate on how the various masks are used.

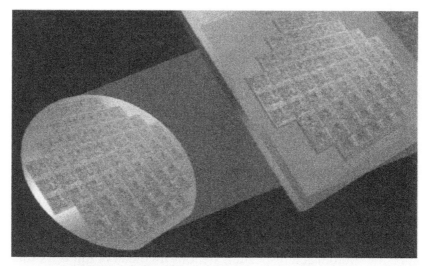

Figure 1-7: Exposing a Wafer Using Photomask

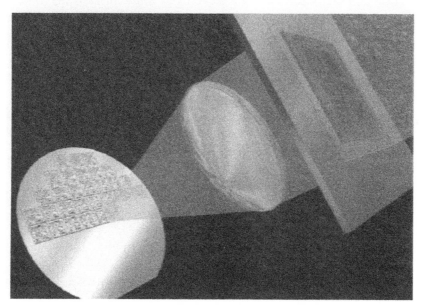

Figure 1-8: Exposing Wafer Using a Reticle

2.4 Process Diagnostics and Metrology

Strict dimensional control is critical to producing state-of-the-art ICs. Many of the devices available today to measure film thicknesses and line widths on the chips did not exist early in the history of the industry. It is no easy matter to measure these tiny dimensions. R&D work is underway for 32 nm technology, that is, a conducting "wire" on the chip that is 32 nm wide. The diameter of a hair is 50,000 to 100,000 nm! Chapter 3, Section 1 has the first of several helpful hints, found throughout the book, regarding the units of measure common to the industry. Here, nm stands for *nanometer* or one-billionth of a meter.

After each critical manufacturing step, the wafers are often inspected. Defects, contamination, processing or operator errors and metrology, and the measurement of critical dimensions (CDs) and film thicknesses, are some of the items that are checked at each inspection. Automated inspection stations do much of the work today.

The dimensions of features on today's chips are so small that even the sophisticated inspection and measurement tools developed in the last twenty years have been rendered at least partially useless. Scanning electron microscopes (SEM) are required to see the smallest features being produced today. Some films are so thin that even classical thin film measuring instruments like the ellipsometer can be inaccurate. Broad spectrum reflectrometry is still useful for film thicknesses, but sophisticated new technologies using ellipsometry are improving monitoring and measurements. New SEM technology is automated and makes the smallest measurements in the fab.

Integrated Circuit Fabrication

The story of building IC's begins now.

3.1 Layering

Semiconductor devices are built in layers. While much of the critical electrical action occurs in the silicon wafer (substrate), a great deal of the chip building process involves the layering of thin films of various sorts on the substrate. Since chips are electrical devices, it is not surprising that layers of electrical conductors are separated from each other by layers of electrical insulators. There is a little more involved in building the whole device, as will be seen later, but that is the general theme of layering.

Thin film deposition is the general name for the technology of adding layers of various materials to the wafer. Layers are also formed by technologies called *oxidation* and *epitaxy*.

3.1.1 Insulators (Dielectrics)

Silicon dioxide (oxide for short), the principle component of glass, is the most widely used insulator in the industry. Other chemicals are sometimes added to oxide to modify its physical characteristics.

Insulators are part of a more general class of materials called *dielectrics*. Dielectrics, as a class of materials, have a variety of properties that help make the chip work better. The text discusses several uses for these materials besides those of electrical importance.

To improve chip speed, special dielectrics are being developed to replace the oxide in some areas of the chip. These new materials are called *low-k* and *high-k* dielectrics. Their use is discussed later in the text.

Early in the wafer fabrication process, high temperature procedures can be used. Simply heating the silicon wafer in an oxygen atmosphere actually grows a coating of thermal oxide all over the wafer; some of the silicon is consumed by reacting with the oxygen to form silicon dioxide. Typically, a temperature of 800° to 1000°C is needed for the oxidation reaction to proceed. The devices being made on the wafer cannot withstand that high of a temperature later on in the process, so lower temperature options must be used.

Chemical vapor deposition (CVD) is one of the processes of depositing a film on the wafer surface. The temperature can be much lower than that needed for thermal oxidation. As another example, some low-k dielectrics are applied in liquid form using a spinning chuck to hold the wafer; deposition is followed by a curing process at 300° to 400°C. Other methods of low-temperature depositions are discussed in the text.

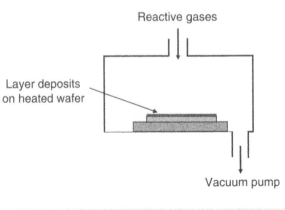

Schematic representation.
There is a wide variety of chamber designs in use.

Figure 1-9: Chemical Vapor Deposition (CVD)

3.1.2 Semiconductors

Silicon technology is the topic of discussion for this book. Silicon is used in two forms: single crystal and polycrystalline (poly). The third form of silicon, amorphous silicon, is not typically used to make CMOS devices, and so it will not be discussed in this text.

The wafer is single crystal silicon. The production of the silicon ingot and the wafers was discussed in Section 2.1. Some chip designs require that the device be built in a thin layer of single crystal silicon that is grown on top of the starting wafer. That type of film is called *epitaxial silicon (epi)* and it is grown in an epitaxial reactor at a high temperature. The epi film is an extension of the underlying silicon crystal.

Silicon can also be deposited on the wafer, usually on top of a film of silicon dioxide. Polycrystalline silicon (poly) forms at temperatures above about 600°C. It is composed of grains of single crystal silicon all stuck together. At lower deposition temperatures, the crystal regularity is lost and the silicon becomes amorphous.

Silicon in an IC almost always needs to conduct fairly well, so it will be doped to improve its conductivity. Doping is discussed in Section 3.3.

3.1.3 Conductors

Connecting the individual electrical components on the chip together to form a circuit is done mostly with metals, notably aluminum, copper and tungsten, but doped poly and silicides are used for some connections. Silicides are chemical compounds that can be made by reacting silicon and certain metals. These silicides conduct quite well and can withstand relatively high process temperatures.

Physical vapor deposition (PVD) is often the method for depositing metal on the wafer. The PVD tool uses large ions attracted to a target by an electric field. The target is made of the material to be deposited on the wafer. The bombarding ions eject atoms from the target. The atoms fly off the target (sputtering) and form a coating on the wafer. Of course, this "blanket" coating will have to be formed into conducting wires, if used for interconnections, as discussed in Section 3.2.

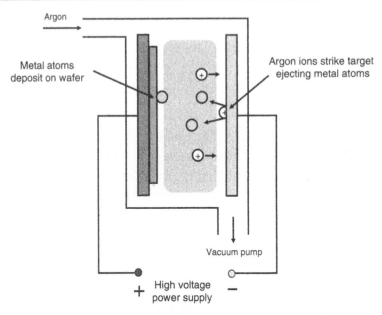

Figure 1-10: Physical Vapor Deposition (PVD)

A popular method of depositing copper is *electroplating*, also called electro-chemical deposition (ECD). The wafer is immersed in a chemical solution containing copper sulphate. A copper plate is also immersed in the solution and when an electric current is passed through the solution between this plate and the wafer, copper is deposited on the wafer.

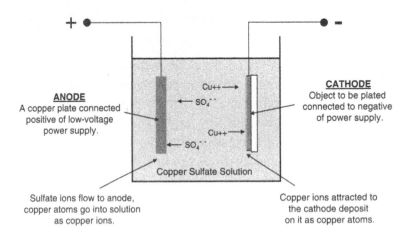

Figure 1-11: Electrochemical Deposition (ECD)

Because of the density of components on the IC, many crisscrossing layers of metal interconnects are needed. Each new layer is insulated from the previous underlying one by using a dielectric material. Holes are cut in the dielectric and filled with metal to make the electrical connections between layers. Eight layers of metal are common today and the number is increasing.

3.1.4 Chemical Mechanical Polishing

Chemical mechanical polishing (CMP) was mentioned earlier as the method used to polish newly produced starting wafers. It also has other uses in making ICs.

The optical tools discussed in Section 3.2.3 used to transfer the pattern from the reticle to the wafer are capable of printing very tiny features on the wafer. But the wafer becomes too bumpy rather early in the process for good pattern transfer; the optical tool cannot focus on all points on the wafer. The wafer must be planarized for best results. CMP is the best technology available for planarizing the surface of the wafer.

New processing methods and the use of copper have also required the use of CMP. The dual-damascene process discussed in the text requires that the copper be removed by polishing. The shallow trench isolation technology also must use CMP to remove excess oxide. Several examples will be discussed.

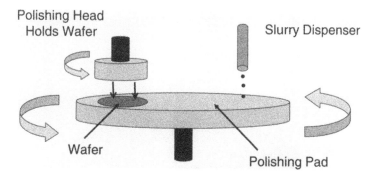

Figure 1-12: Chemical Mechanical Polishing (CMP)

3.2 Patterning

The patterning process transfers the pattern from the reticle (or photomask) to the wafer using photolithographic technology. A photoresist template, also commonly called a mask, just to confuse everyone, is produced on the wafer surface that is a duplicate of the pattern on the reticle. One of two functions is performed by this template: it is used to protect parts of the wafer from chemical attack during an etching process or to block the implantation of dopants from the covered areas. The use of photolithographic techniques is also discussed in Section 2.3.

3.2.1 Photoresist Coat

Photoresist (resist) is a light-sensitive plastic material that is first dispensed in the form of a solution. The wafer is positioned on a holder or "chuck" that holds it in place with a vacuum while the chuck spins. The resist is applied to the wafer by dispensing a small amount of the thick, sticky liquid onto the center of the wafer. The spinning throws off excess resist and helps to determine the thickness of the resist film; a higher spin speed will drive off more resist and leave a thinner film behind. Other factors, such as the viscosity of the resist, contribute strongly to the film thickness.

A soft bake at about 100°C drives off most of the solvent. The resist is stabilized (solidified) by the bake and the photosensitivity and solubility in developer solution are influenced, too.

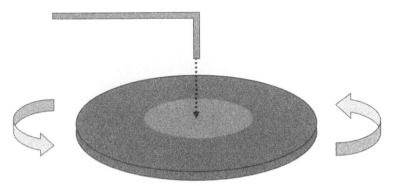

Thin layer of photoresist solution spun on the wafer.

Figure 1-13: Photoresist Dispense

3.2.2 Exposure

Most exposure operations are done in a stepper. This tool holds the reticle and projects light through it. The resist is sensitive to ultraviolet (UV) light. The areas exposed to UV will dissolve in the developer solution. The unexposed areas are left behind and form a template or mask on the wafer.

The name "stepper" is short for the formal name of the technology: step-and-repeat reduction-projection printing. Step-and-scan is also a common technology found in the fab today. The image on the reticle is reduced in size by lenses. Most reduction systems in use today are 4x—that is, they reduce the image to 1/4 of its original size on the reticle.

The step-and-repeat or step-and-scan describes how the tool exposes the wafer. Since the image is reduced in size, only a small portion of the wafer can be exposed at a time. The wafer must be moved or stepped after each exposure. The wafer is exposed "step-wise" until the pattern has been transferred to the whole usable surface.

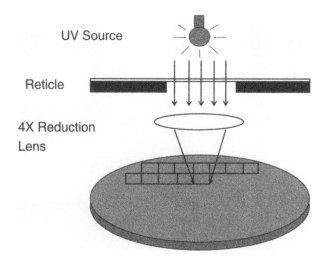

UV Source

Reticle

4X Reduction
Lens

Figure 1-14: Photoresist Exposure Using Stepper

3.2.3 Develop

The exposed wafer is positioned on another spinning vacuum chuck, but this time a developer solution is sprayed on it. The exposed resist dissolves leaving the resist template (mask) behind.

3.2.4 Pattern Inspection

If an error in processing has created any imperfections in the photoresist image it can be stripped off and reworked. Now is the time to find any problems because the next processing steps will permanently transfer the pattern to the wafer. Imperfections could result in scrapping many wafers.

Automated inspection tools check the pattern for defects and measure important feature sizes called *critical dimensions* to monitor the quality of the product.

3.2.5 Etch

Most of the masking steps are followed by an etching step. Plasma etch is the dominant technology for permanently transferring the resist mask into the film or films below. The wafer enters an evacuated chamber. A gas is introduced and electrically charged (ionized) to change it into a plasma. The plasma is highly reactive and allows better control of the feature shapes and dimensions than the wet etch method.

Plasma Etchers

Technically, plasma etch reactors only make a small amount of the injected gas into a plasma. Typically, less than 1% of the gas particles are ionized at any one time. However, as discussed in Chapter 4, that is all that is needed to get the job done.

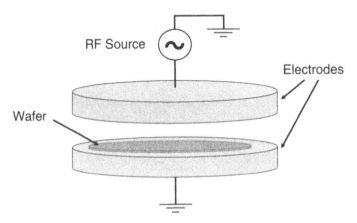

Figure 1-15: Plasma Etch

3.2.6 Implant

Some steps in the process use the resist mask as a protective template to permit dopant species to enter only selected areas. This implant mask is needed when an implanter is used to dope the wafer. An implanter is a particle beam machine that accelerates dopant ions and drives them into the wafer.

It is also common to etch the required implant masking template into a protective film (often a dielectric) that is deposited on top of the material to be doped. In this case, the process looks just like any other photo/etch step with the wafer going to implant after the etch sequence.

Section 3.3 will cover more on doping.

3.2.7 Photoresist Strip

When the resist mask is no longer needed it is removed, often by etching it off in an oxygen plasma (the resist is simply burned off) or through the use of a liquid stripper.

3.2.8 Etch Inspect

The features that were just produced on the wafer are inspected and critical dimensions measured. Defects found at this point often require the affected wafers to be scrapped.

The wafer is now ready to have another thin film deposited so it can return to patterning for the completion of another layer. The wafer loops through these steps until the device is complete.

3.3 Doping

Dopants are introduced to alter the electrical conductivity of the doped region. They can be added to the wafer in a variety of ways. The dopant is added to the silicon melt at crystal growth so the silicon ingot is either n- or p-type. The chemistry used to produce epitaxial silicon will usually include dopant species. The wafer can be doped simply by putting it in a furnace and introducing the dopant in gas form.

Today, almost all of the doping of state-of-the-art ICs is done by implanting the dopant using a particle-beam machine called an ion implanter. As mentioned above, the implanter accelerates ions down a long tube and drives them into the wafer. The dopant enters the wafer wherever the implant mask has an opening. This method allows great precision in the placement of dopant as well as its concentration, or "dose." It is the doped regions that form critical sections of MOS transistors that will be discussed in detail in the text.

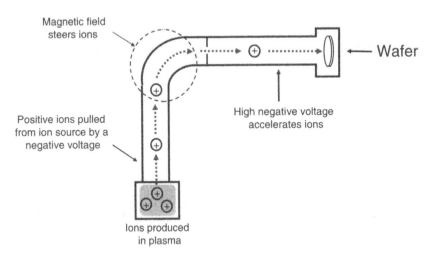

Figure 1-16: Ion Implantation Doping

3.4 Process Control and In-line Monitoring

Production lines must be constantly monitored in an attempt to prevent costly misprocessing. Statistical process control (SPC) is the best method for real-time monitoring of a manufacturing line. SPC charts are updated continually as each inspection is completed. The chart provides signals of a drifting process and often prevents expensive production errors.

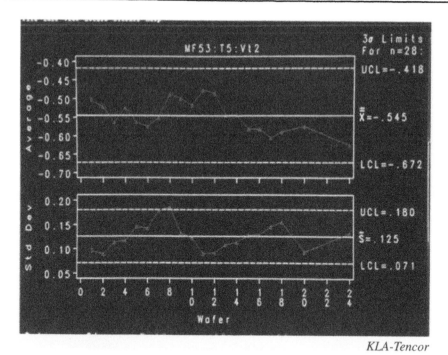

KLA-Tencor

Figure 1-17: Statistical Process Control Chart

Test and Assembly

The completed wafers are ready to become individual packaged parts if they pass electrical tests.

4.1 Electrical Tests

Two types of electrical tests are done while the wafers are still whole. First, several dozen basic measurements, called parametric tests, are made on individual transistors and other test structures in test patterns on the wafers. This is the first check on proper fabrication.

The second test is a circuit function test. Each chip is tested with a probe that has many tiny electrodes. A computerized system performs many tests on the chip and collects statistical data on the product. The test system keeps a map of passed and failed die on every wafer. Failed die are often marked with an ink dot or "inked out."

4.2 Die Separation

The chips are also called die (discussed in Chapter 8). The tested wafers are "diced" by putting each wafer on an adhesive holder and cutting the chips apart with a diamond saw.

4.3 Die Attach and Wire Bonding

The good die are removed from the adhesive and attached to a chip package. The dual inline package is the most common and contains a lead frame to which the chip is attached. The lead frame is a mounting that has the appropriate number and configuration of contacts or "leads" to attach to other electrical devices, circuit boards and the like.

The chip itself is attached to the lead frame using special solder such as gold-germanium alloy or epoxy. Then the tiny gold or aluminum wires that connect to the bonding pads on the chip are attached using special techniques that do not damage the pads.

4.4 Encapsulation

Encapsulation is the process of putting the chips into protective packages. Many packages are available for chips. The intended use of the part is key in deciding how to encapsulate it. Military hardware is exposed to harsh environments so tough packaging is used. Simple plastic packages, such as the dual inline package, suffice for many chips. Metal, ceramic and composite packaging material are all used.

4.5 Final Test

The packaged parts receive a functional test similar to that given earlier to the die on the wafers, but it is usually more extensive since many products cannot be tested at full speed or power until final assembly is complete.

The parts are now ready for shipping to stores or customers.

Summary

Chapter 1 summarized the components of IC production. Chapter 2 begins the story of building a chip from the ground up. In Chapters 3 through 7, the reader will actually build a chip, figuratively speaking, taking the wafers through the entire process, watching each component of the chip come together.

CHAPTER

2

Support Technologies

Introduction

In Chapter 2, you will learn:

- What are support technologies?
- How the manufacturing environment is kept clean
- How silicon wafers are made
- How an integrated circuit is designed
- How the chip design is transferred to the manufacturing line

Definition: *Support Technologies* are the ancillary or support operations that are critical to the integrated circuit manufacturing process.

> ### Chapter 2 Selection of Topics
>
> To simplify the organization of this text, some topics are included in the support technologies chapter even though they are not technically providing support to the manufacturing process but are, rather, a principal part of it. The reason for that choice is to reinforce the focus of this book, which is the actual fabrication of the integrated circuits. So even though the chip could never be manufactured without first being designed, the discussion of the design process is found in Chapter 2.

The principal support technologies to be addressed are as follows:

1. Contamination control
2. Silicon wafer manufacturing
3. Circuit design
4. Photomask/reticle manufacturing

Contamination control will begin the discussion of this chapter. The chipmaking environment is so sensitive to contamination that even the machinery used in the process must manufactured in a cleanroom. The wafers that are manufactured from silicon ingots must be extremely clean before they can enter the cleanroom. Even the chemicals used in the process must be of the highest possible purity and be nearly free of particles.

Manufacturing the wafers is a highly sophisticated process. State-of-the-art fabs use 300 mm (12 inch) diameter wafers—as big as a dinner plate. The ingots of silicon produced today are often one meter long and weigh close to 165 kg (360 pounds).

IC design is the only operation discussed here that does not need to be done in a cleanroom. But the photomasks produced from the design are made in an ultra-clean environment.

Another topic of importance to the building of ICs is process diagnostics introduced in Chapter 1. Critical dimension measurements, film thicknesses, defect identification, particle counting and other measurements are indispensable to the process. There are many, many technologies that are put to work to do the job—so many, in fact, that to do justice to the topic would lead the discussion far away from the focus of this book. The reader is invited to read about metrology separately. The bibliography contains many helpful references that will serve as a starting point.

Contamination Control

2.1 Why Control Contamination?

One of the most amazing things about state-of-the-art ICs is how small the features are on the chip. Features include conducting lines, holes made in insulators and other structures that are easily identifiable under a microscope. It is becoming very difficult to make meaningful size comparisons. The preferred example that has been used for years is the diameter of a human hair. Typically, a fine hair is about 50 microns (micrometers) in diameter, and a coarse hair will be about 100 microns in diameter. When the conducting wires or other features on the chip were one micron wide, or even one-half micron wide, one could still get oriented.

The feature sizes in the chips are dropping below one-tenth of a micron. Optical microscopes are heavily relied upon in the fab but when feature sizes drop much

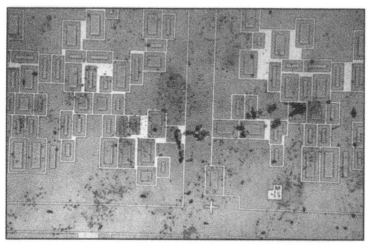

Figure 2-1: Wafer exposed to Air Outside Cleanroom

below one micron their usefulness is limited. So the smallest features on the chip are effectively invisible to an optical microscope. The scanning electron microscope (SEM) is quite common because it is indispensable in today's manufacturing environment. A SEM is needed to actually see the smallest features on the chip.

Here is an example: 65 nm technology can fit 10 million transistors in the space that is roughly the size of the tip of a ballpoint pen. It seems that technology has exceeded the ability to connect it to the "real" world.

Tiny Dimensions

It is becoming very difficult to relate the feature sizes on a chip to everyday experience. A human egg is 100 microns across—huge compared to a feature on a state-of-the-art chip. Even a sperm is comparatively large, about 5 microns in diameter at the head. A typical bacterium is 0.5 microns in diameter. That's still awfully large, but getting close. Shifting to virus-sized dimensions in the few nanometers range (1000ths of a micron) clearly defeats the purpose; such small things are far removed from the realm of our everyday experience.

Why are these tiny features on the chip so vulnerable to contaminants? Simple particles of dirt that are smaller than the smallest feature on the chip will cause trouble. They can cause a portion of the chip to malfunction by causing the feature to be misshapen, and therefore, not function as needed. Some types of particles will block the formation of patterns on the wafer by interfering with the photolithography process; still other particles can cause a conducting path to form between conducting lines on the chip, shorting them out. The following illustration shows an example of damaged areas on a chip.

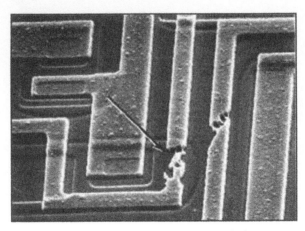

Figure 2-2: Pattern Defect (*Semiconductor Picture Dictionary*—see Bibliography)

Chemical contaminants that corrode metal structures or dissolve thin film materials are a constant hazard. These chemicals are often the very ones used in the manufacturing process.

Even smaller contaminants are dangerous. Mobile ion contaminants are ions of metals that can migrate to the heart of electrical devices such as transistors and will damage or destroy them.

With critical dimensions on ICs as small as they are today, wafers must be kept in a contamination-free environment throughout manufacturing.

2.2 Contamination Sources

For many years, the people working in the fab were identified as the principle source of particulate contamination in the cleanroom. Everyone continually sheds microscopic particles of dry, dead skin and hair. Dust and dirt is also carried on clothes and in hair. Makeup is especially dangerous because it not only adds to the particulate shedding, it also contains chemical contaminants that will ruin the chips. Makeup has been forbidden inside the fab since the mid-1970s.

The machinery used to make the chips is the main particle source in state-of-the-art fabs today. Maintenance procedures are critical to keeping the tools free from contamination. Mechanical devices with moving parts suffer wear; lubricants outgas; chemical vapors condense on the interior walls of the chambers and loadlocks and can build up until they flake off and drop on the wafers.

Significant improvements continue to be made in processing tools. For example, self-cleaning processing chambers are offered for some types of reactors. In other cases, quick-change replacement chambers and liners are available.

Fabrication will always be a part of the discussion of contamination. The process produces particulates and chemical contaminants. For example, chemical vapor deposition (CVD) and other thin film deposition processes will often deposit films on the interior walls of the reactor as well as on the wafer. If allowed to build up, these materials often break away and fall on the wafer, and then an unwanted particle could be incorporated into a newly deposited film.

Chemical mechanical polishing (CMP) is a good example of a processing operation that creates particles. CMP uses tiny abrasive particles that are delivered to the wafer surface in a liquid slurry. The abrasive grinds off tiny particles from the wafer surface. All of these particles must be removed at the end of the processing step or the wafer could be ruined.

Processing chemicals are an important consideration. Residues left behind by processing liquids and gases may later cause damage such as corrosion of interconnecting metal wires on the wafer. Clearly, all unwanted chemicals must be removed from the wafer after processing.

Deionized water (DI water) must be used at all stages of manufacture. The ions (see Appendix A, Science Overview) in tap water would transfer to the chips and ruin the electrical devices on it.

2.3 The Cleanroom

The fabrication (fab) area where chips are manufactured is called a *cleanroom*. The cleanroom accomplishes a large part of the task of keeping the wafers free of contamination.

Cleanrooms are classified according to the number of 0.5 micron or larger particles that are present in a cubic foot of air. State-of-the-art fabs are Class 1 cleanrooms. That is, they have only one particle that is 0.5 microns or larger in each cubic foot of air in the fab. Class 10 and Class 100 cleanrooms are used for less demanding processes.

Figure 2-3: Inside the Cleanroom (Intel)

The cleanliness of the fab is maintained by HEPA or ULPA filters. The cleanroom is designed to circulate the air in such a way as to optimally carry away any airborne particles. The airflow leaves the HEPA filters in the ceiling and travels straight downward to the floor in a laminar fashion. The vertical laminar flow (VLF) prevents particles from being scattered sideways and ensures that they are carried away. A typical circulation path is shown in Figure 2-4.

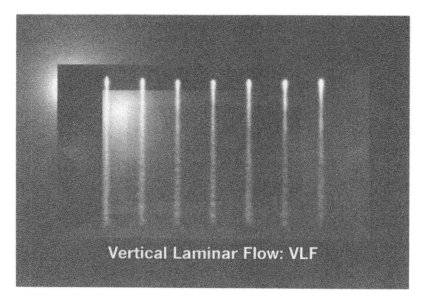

Vertical Laminar Flow: VLF

Figure 2-4: Laminar Flow Air Pattern

Air coming into the fabrication area from outside goes through a processing plant to filter it and to set the temperature and humidity levels to match the air requirements inside. It was not unheard of in year's past for a sudden change in the weather to cause some processing problems in the fab because the machinery controlling the environment could not react fast enough and would become temporarily overloaded. State-of-the-art fabs built today are over-engineered to compensate for that eventuality.

The cleanroom coverall, fondly known as the "bunny suit," has evolved over the years to resemble a space suit. Today's suits are made of space-age materials, such as Dupont's Kevlar®, and are equipped with filters and fans to draw air through the suit but not let any particles get out.

Gowning procedures are a critical part of the training for everyone entering the fab. The gowning room and gowning procedure allows a person to "gown up" without transferring too many particles to the outside of the bunny suit. Everyone entering the fab must then pass through an air shower, which is a small room with air vents blowing air from all angles in an attempt to remove all loose particles from the outside of the suits.

Historically, the fabrication area has been laid out in bays branching from a central spine. Today, there are new layouts being introduced that consist of a large work area for each major process step. The newer multifunction processing tool designs, or cluster-tools, discussed in later chapters facilitate this innovation.

Wafers are always held in cassettes, and the cassettes are placed in carrying boxes. State-of-the-art cleanrooms move the wafers from one area to another within the fab using transport systems called *automated material handling systems (AMHS)*. Some, like the monorail-loop transport system are positioned near the ceiling and are loaded by placing the carrying boxes in an elevator that automatically places the box on a "car." The cars travel on tracks along the walls overhead. Other transporters are closer to tabletop level, such as the automatic guided vehicle system.

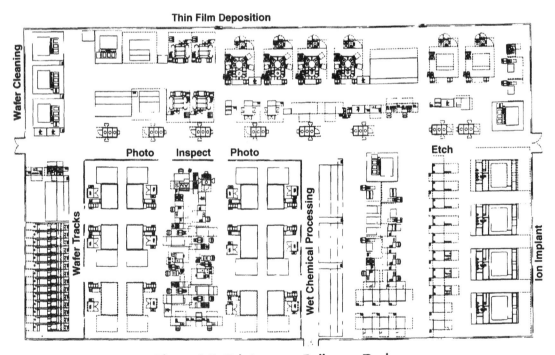

Figure 2-5: Fab Layout—Ballroom Design

Minimizing wafer handling by humans is important. Fabs can be highly automated such that only a few people are ever needed to move a box of wafers. Processing equipment is manufactured in clustered configurations; when possible, the wafers are put under vacuum and kept there as long as possible, thus removing the risk of atmospheric contamination.

Recordkeeping, note taking, lot tracking and other paper-intensive activities have always been a problem in the cleanroom. Paper sheds a large amount of particles and cannot be used in the cleanroom. Pencils and pens also produce particles. Lint-free paper was developed from polymer fibers and it helped solve the problem, as did special cleanroom pens. Today's fabs will not tolerate even the very low levels of particles produced by the lint-free paper. Computer terminals, spaced every five or ten feet apart throughout the facility have replaced writing implements and paper.

Some fabs have achieved low particulate levels without making the entire facility into a Class 1 room. It is possible to use mini-environments that surround each processing tool and protective boxes to transport the wafers. The industry uses a standard mechanical interface (SMIF) to ensure compatibility with boxes, loading systems and processing tools. In these fabs, the only areas that require strict gowning procedures are small rooms where wafer inspections are done by fab personnel rather than automated inspection tools. Of course, some cleanroom procedures are still necessary for the areas with higher particulate classifications.

Many other issues such as chemical purity, point-of-use filters for gases, production of deionized water and the elimination of static electric charge are a few of the topics that are beyond the scope of this discussion. The reader will find references in the bibliography that will lead to further reading about cleanroom technology.

SECTION 3

Crystal Growth and Wafer Preparation

3.1 Introduction

Several different semiconducting materials are used for electronic device applications. Chapter 1 mentions the most prominent elemental and compound semiconductors favored for production. Because silicon technology represents the vast majority of commercial IC fabrication, it is the focus of this section.

In this section, the purification of silicon, single-crystal silicon growth and wafer manufacturing is described. The following techniques are the ones that dominate the industry today.

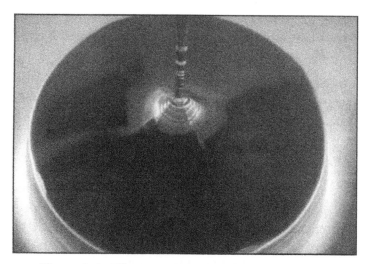

Figure 2-6: Silicon Ingot Being Pulled from Melt

Process Flow:

1. Silicon Purification
   1.1. Metallurgical Grade Silicon
   1.2. Electronics Grade Silicon

2. Czochralski (CZ) Crystal Growth
   2.1. Doping

3. Ingot Shaping
   3.1. Notch Grinding

4. Sawing Wafers
   4.1. Lapping and Grinding
   4.2. Edge Grinding and Smoothing
   4.3. Chemical Etch

5. Chemical Mechanical Polishing
   5.1. Wet Clean

6. Laser Inspect

7. Ship to Customer

This process flow looks quite detailed, but the discussion will summarize each point briefly and emphasize the purpose of the most important steps.

Definition: A *Substrate* is a foundation or basis. It is another name for the silicon wafer. The integrated circuit is built partly within and partly upon the semiconductor substrate.

3.2 Silicon Purification

The production of single crystal silicon wafers with the chemical consistency and crystalline structural properties required to make ICs requires that the silicon be refined to an unprecedented level of purity. The ingot must be made from silicon that has impurities in the parts-per-billion-atoms (ppba) range. Silicon of this purity is called *electronics grade silicon (EGS)* and is one of the purest materials ever made. It is 99.999999999% pure!

Chemical Purity

Material with a purity in the parts-per-billion-atoms (ppba) range still contains impurities at levels of ten million-million atoms per cubic centimeter of silicon ($10^{13}/cm^3$). Even though this silicon is the purest material commonly produced in the world today, it is still far from perfect. Many strategies are employed in the design and manufacturing process to compensate for the inability to remove 100% of all impurities from these materials. It seems amazing that ICs work at all, but the sophisticated tricks of the trade are even more amazing. Many of these secrets will be revealed in the discussions to follow.

The *starting material* is a type of sand called *silica,* which is made of silicon dioxide with very low impurity levels. Silica is actually another name for silicon dioxide. The silica is reacted with carbon, producing metallurgical grade silicon (MGS) at a purity level of about 98%.

Improving the purity of the silicon is no easy task as proven by the complicated processing steps to follow. The solid MGS is converted to a liquid chemical called *trichlorosilane (SiHCl$_3$).* In this form, it can more easily be purified; it is refined in a very involved distillation process that is beyond the scope of this text. The purified trichlorosilane is then reacted with hydrogen in a chemical vapor deposition (CVD) reactor that extracts the highly purified solid electronics grade silicon. CVD is discussed at length later in the book.

The EGS just produced is polycrystalline in structure. ICs must be built on single-crystal silicon. The next step produces single-crystal silicon.

3.3 Czochralski Silicon Growth

The Czochralski (CZ) method of silicon growth is named after the man who invented it, Jan Czochralski. Nearly all of the semiconductor manufacturing industry's silicon wafers are cut from single crystals grown using this method.

In the CZ method, a single crystal of silicon is grown from silicon melted in a quartz-glass crucible. This ingot of silicon is quite large and is produced in a two-story high furnace. State-of-the-art fabs use 300 mm diameter wafers (about 12 inches) so the ingot can be quite large.

The process begins with a quartz-glass crucible filled with chunks of polycrystalline silicon. The crucible is heated in the crystal grower until the silicon melts. A temperature of about 1414°C (white heat) must be reached to melt the silicon. A seed

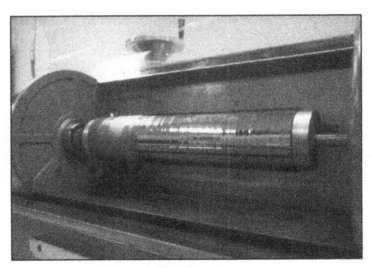

Figure 2-7: Silicon Ingot

crystal having the desired crystal orientation is then lowered until it touches the top of the melt. Many subtleties exist in the successful production of an ingot, including the rotation of the melt and the seed in opposite directions. The seed crystal acts as a template, providing the structural pattern for the crystal. The molten silicon freezes onto the seed as it is withdrawn. The entire ingot is a single crystal of silicon matching the atomic arrangement of the seed crystal. The process takes more than an hour for each inch of ingot length.

Crystal Orientation

Every crystal can be described as a regular arrangement of many unit cells. The unit cell is the smallest unit of a crystal that still demonstrates the regular structure of atoms found throughout the crystal.

Some unit cells, and therefore, single crystals, look the same when viewed along any principal axis. Silicon, however, does not. The orientation of the silicon crystal is selected based upon the electrical characteristics desired in the chip. A system of orthogonal axes is used to designate the crystal orientation. The bibliography includes several texts with a more complete explanation of this characteristic of a crystal and the notational system used for each orientation.

For example, a specific type of IC is manufactured on [111] silicon because it etches faster in the direction of that plane in certain etch solutions, but more slowly along the other two axes. As a result, a precisely shaped trench can easily be etched into the silicon, forming an important structure needed to make the chip.

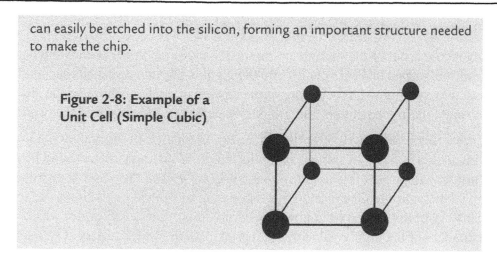

Figure 2-8: Example of a Unit Cell (Simple Cubic)

The vast majority of silicon produced for chip manufacturing is doped in the melt. Crystals are grown with a wide range of doping levels. A chipmaker's choice of doping level and species is dictated by the nature of the product to be fabricated. The starting wafers needed to fabricate high voltage rectifier chips are much different from those needed for a CMOS IC. CMOS chipmakers choose doping levels in starting silicon corresponding to their process design and electrical parameter goals.

The role that the doped substrate silicon plays in the device being manufactured, as well as doping other films used in the process will be discussed in the coming chapters. The topic of semiconductor doping is also discussed in Appendix A, Science Overview.

Another impurity, oxygen, is unintentionally incorporated into the silicon. The quartz-glass crucible is made of high purity silicon dioxide. Some of the crucible reacts with the silicon in the melt, adding a small amount of oxygen to the single crystal. The oxygen dissolves in the silicon (a solid solution) and will precipitate out and react with the silicon during some of the high temperature processing steps, forming silicon dioxide (oxide). It turns out that the oxide is a big help in controlling contamination. The silicon dioxide tends to clump up, forming dislocations in the silicon crystal. These imperfections are gettering sites (see Appendix A) that will trap and hold some types of contaminants.

3.4 Shaping, Grinding, Cutting and Polishing

The ingot comes out of the melt as an irregularly shaped cylinder with pointed ends. The teardrop-shaped ends are cut off and the ingot is put into a special machine that grinds it into a smooth-sided cylinder of the desired diameter. Naturally, the silicon is hard so diamond-tipped cutters and grinders must be used for much of this work.

Since a round wafer will offer no feature for orientation of the wafer on a wafer-handling tool, a notch is machined along the length of the ingot. It will soon become clear just how important it is to know which way is "up" on the wafer at every stage of chip manufacturing. One or more flat sides were machined on the ingot earlier in the history of the industry but now a small notch is cut on every wafer which consumes less of the highly valuable "real estate" on the wafer surface. Obviously, manufacturers want to produce as many chips per wafer as possible.

Next, the wafers are made by sawing the ingot with special abrasive-coated wires. Stainless-steel blades with diamond edges are often used for cutting wafers smaller than 300 mm.

The wafer surfaces are fairly rough at this point. They move to the lapping and grinding tool where they are further processed until they satisfy a strict thickness specification. The lapping and grinding tool accomplishes this through the use of two counter-rotating plates and a slurry with an abrasive in it. The wafers are placed between the two plates, and the slurry removes the excess silicon.

Wafer edges are an important aspect of wafer production. If the edges of the wafers are rough, they shed tiny particles of silicon that can damage or ruin the chips. Crystal defects are also an issue if the surface of the wafer edge is not well polished, so the edges of the wafers are rounded with a grinding tool and then highly polished.

The wafer-shaping processes used up to this point have brought the wafer to an intermediate stage as far as general size and shape. Next, surface damage and contamination must be removed. The first step in this phase of the process is a wet chemical etch using a combination of acids, usually including hydrofluoric acid and nitric acid. Contaminants from the processing tools and particles from lapping, grinding and polishing are removed.

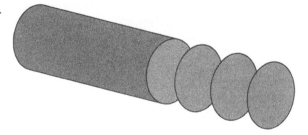

Figure 2-9: Silicon Ingot Ground and Sliced into Wafers

Chemical mechanical polishing (CMP) is a major part of chipmaking today, and applications appear often in the coming chapters. However, the first application of CMP in the industry was polishing wafers. It is at this point that the side of the wafer on which the devices will be built is brought to a highly polished finish.

Why should a high shine have any effect on making the ICs? The surface roughness must be dimensionally smaller than the size of the features being produced on the wafer. The electronic components of the chip are partly built into a very thin portion of the upper surface of the wafer. There is always a high likelihood of defects on the surface of a crystal—right where the devices will be built. In order to minimize the number and severity of crystal defects, the surface is made as perfect as is humanly possible.

CMP strongly resembles the lapping and grinding process discussed earlier. The CMP slurry is specifically designed for this fine polishing process and uses carefully selected abrasives and liquid components. The liquids in slurries may contain acids, oxidizers or bases in a proprietary mixture. The polishing pad is also a specially designed material.

The post-CMP clean is very important. The process of polishing a surface removes very tiny particles of the surface material. The slurry contains tiny particles and they must be removed or the chips will be damaged. The chemical contaminants from the slurry must also be removed.

3.5 Final Inspection and Shipping

A laser inspection is done prior to shipping the finished wafers. Special machines use lasers to inspect the surface of the wafers for all types of defects. These automated tools identify defects quickly and accurately.

Flatness, bowing, resistivity, physical measurement and many other specifications are checked by automated tools.

> **Wafer Flatness**
>
> Wafers satisfy some of the tightest specifications known to man. If an airport runway was as flat as a wafer, its surface would vary by one inch or less per mile!

When the wafers pass final inspection they are placed in cassettes, wrapped in layers of plastic and shipped to the customer.

Circuit Design

4.1 Introduction

The integrated circuit (IC) is simply a lot of discrete electronic devices like transistors, diodes, resistors and capacitors that are all built on a single piece of semiconductor material and wired together; these devices are "integrated" into a single chip. The process of designing an IC is quite complex. Today's chips may contain well over one billion of these discrete devices. To call the design process "nontrivial" is something of an understatement. It is a highly sophisticated procedure utilizing some of the most advanced software tools ever devised.

Comparatively simple ICs can be designed in several months. A product like the radio frequency identification (RFID) chip that performs only one simple function is an example. Microprocessors are a very different story; they take years to design. Microprocessors are complex chips that perform many different functions on the same chip. It takes several specialized design groups to get the job done. Rigorous verification of functionality alone takes months.

Miniaturization is the name of the game in chipmaking. Planning the shrinkage of the size of the part adds to the design time. Designers must coordinate with manufacturing to ensure that it is possible to make the chip that they have designed. The space program drove the early development of the industry. Now Moore's Law is mistaken for a law of nature because it has proven to be so reliable.

> ### Moore's Law
>
> Moore's Law is named after Gordon Moore, one of the founders of Intel. In the mid-seventies, he observed that manufacturers of computer memory chips were able to double the number of transistors in their products every eighteen months. Actually, his first approximation was twelve months, but he soon modified it to the eighteen month figure that has been used ever since. The corresponding exponential rise in chip performance is largely responsible for the growth in computer performance without a corresponding increase in price.
>
> Moore's observation has proven to be incredibly accurate, and no one was more surprised than Moore himself! Many people expect to reach the end of its validity any day now, but the innovators in the industry continue to produce ever more complex ICs with more and more transistors at that amazingly consistent rate.

Several kinds of companies design ICs. Independent design firms are probably the most numerous today. These companies do only the design process and contract out to a *foundry* for the production of the chip. Some very large firms such as Altera and Xilinx are so-called fabless companies. Fabless design houses use *Design Rule Sets*, generalized instructions, guidelines and restrictions on the IC design, as provided by a particular foundry or a group of foundries. These design rules tell the designers exactly what the manufacturing capabilities are for the foundry so that they do not design a chip that cannot be produced by that fab.

Foundries are becoming the backbone of semiconductor manufacturing. The foundry manufactures chips designed by other firms. TSMC and UMC in Taiwan are examples of firms that do only foundry work. Motorola, IBM and Samsung are a few well-known manufacturers that do foundry work for the independent design houses, as well as making chips designed in-house.

The independent device manufacturer (IDM) is the firm that only makes chips in its fab that are designed by its own engineers. While these types of companies once dominated the chip-making business, IDMs are becoming rare today because of the high cost of building a new fab. Most firms with a manufacturing facility will entertain offers to do foundry work.

4.2 Product Definition and New Product Plan

Definition: A *Schematic* is a symbolic representation of an electrical circuit. The drawing may include one or more transistors, capacitors, diodes, power supplies or other electrical components, all wired together.

Definition*: Design Rules* are a set of specifications that describe in detail the size of the chip, the size of the features on the chip, the spacing of features and components, electrical parameters and the steps to be followed in the design process to ensure that it runs smoothly and that all checks and verifications are performed.

There are thousands of different ICs made today, but the need for new chips with better or unique designs is as strong as ever. It only takes a good idea to start a new chip-design company. A good idea with a good business plan will attract the needed venture capital. Many of these startups are found in "Silicon Valley" and other innovation centers around the world.

Frequently, the new chip designs are motivated by a customer need. A chip-design company's marketing and sales force will bring requests for a new product from the customer. Also common are innovations within an in-house design group that can add a new line to an existing one or even develop a new part with a perceived need. Other innovations are the result of research or R&D work that develops a new technology to be used for making ICs.

No matter how great the idea for a startup or new chip design, electrical engineers and product designers must be interested in using the new part in their end products. In other words, there must be a market for the new product. Applicability, availability and reliability must be proven for every new chip entering the market.

The process of designing a semiconductor chip starts with the Product Definition. With a clearly defined product concept, a New Product Plan (NPP) is created by a group of key players. It is the master plan for the chip. The NPP often originates with marketing and may be adjusted by designers if the specification cannot be met.

Definition*: The New Product Plan (NPP)* is the master plan for the chip. It contains requirements from marketing, business reliability, technology, operations (fabrication) and engineering (chip design). This information flows to block plans, chip planning and all other downstream chip designing activities.

Some topics in the NPP
architecture
density
speed
voltage
power consumption
package support
die size estimate
bonding options
business opportunities
staffing requirements
estimated schedule
cost analysis
defect density requirements

For example, the marketing department continually interfaces with customers. They keep their fingers on the pulse of the market to determine when the need for a new product appears. Do our customers need analog chips for stereos? Do they need memory chips to store information in their product? Do they need a microcontroller that can be programmed through software? What if the military needs a chip that will work in space on a satellite? Based on its market analysis, the marketing department then creates a product definition; a chip design that will fill a perceived need, and therefore will sell well. Marketing engineers will then work with the design engineers to produce the new product plan.

Perhaps marketing receives a request for a new chip from a customer. Since the chip must interface with other devices in the customer's end product, the customer's requirements will be integral to the creation of the NPP.

New product plans are, by necessity, flexible. Regular meetings of all interested parties are held to review the progress of the design and to discuss the inevitable problems and revisions. Design, marketing and customer representatives all participate, as appropriate.

After the new product plan has been decided upon, the chip *specifications* can be written. These specifications are official documents written by the design and logic engineers of a design team. Many of these particulars come directly from the NPP. They detail the parameters of many intricate circuit characteristics such as voltages, current requirements, power consumption, gain and speed. This, in turn, becomes the basis for the design of the circuit schematics, as well as selection of a foundry's technology.

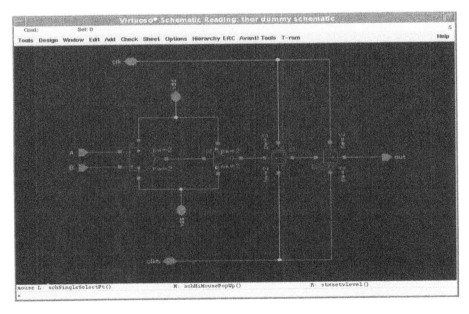

Figure 2-10: Example of a Schematic

Military products are more difficult to design because the ICs must work under a broader range of temperature and environmental conditions than commercial products. The "mil-spec," as it is known, requires the chips to stand up to extensive reliability testing—rigorous thermal shock tests, physical shocks and sometimes even radiation exposures that could come from nuclear weapons. They are also more costly to produce because the government requires more paperwork and more stringent process controls.

There are three basic types of design specifications:

1. *Customer Design Specifications* include all of the requirements from the customer's side. What are the input/output voltages? Are there frequency limitations? What type of signal is being processed? How will the chip interface with the existing product? What package best suits the customer's end product? For example, commodity products such as memory chips or analog-to-digital chips have a range of standard packages for customers to choose from.

2. *Procedural Design Specifications* specify the procedures to be followed during the design process. Are cross-functional groups to be used? Will teams of engineers work on separate segments of the chip or will a single engineer be responsible for each section? How will each piece of the design be passed on upon completion? Procedural design specs also let the team know when a step in the procedure is done correctly. These specifications can be extremely detailed and enumerate how many times a schematic or drawing can be revised or if the author's name must be on the document. These types of specifications are required when a large team of engineers are working on a complex design.

3. *Dimension Design Specifications* enumerate the sizes of the chip and all the components on the chip—the critical dimensions of the transistors, the size of the memory blocks, the size of the chip itself, the width of conducting metal lines and so forth. They are commonly known as *design rules*. The dimensions of the chip and its components often must be adjusted during design, but any large departures from the original plan can strongly affect the value of the chip to the end user.

4.3 The Design Team

A lot of terminology is included in this section that goes beyond the familiar and is offered as a "teaser" to encourage the reader to explore the field in more depth.

Many different types of engineers are needed for the design team.

The most well known engineer is the *design engineer*. This specialist holds an electrical engineering degree coupled with the expertise and creativity to devise new ways to make circuits work. The design engineer utilizes his or her skills to create a schematic that will not only perform a specific function, but do so in a unique way. Creating a unique design makes it eligible for a patent, thereby protecting the company's intellectual property. Over the years, design engineers have transitioned from using paper and pencil to draw schematics, to using sophisticated software that models the final behavior of the circuit immediately after it is drawn. The schematic becomes the basis for many of the design process tasks; circuit simulation and functionality testing, the blueprint and instruction set for the layout engineer and the basis for the documentation that must accompany every new electronic device.

The *layout engineer or layout designer* does the artwork that turns the schematic into a drawing of what the entire circuit will actually look like on the chip. The layout engineer is responsible for planning the shape of *functional blocks:* circuit designs are created in smaller chunks and later assembled into the finished IC. The layout group plans the wiring locations, transistor connections, and conducts verification of the design rules and connections.

These circuit diagrams are converted into layers by the design software. Integrated circuits are built one layer at a time, so each layer is uniquely identified in the drawing program so it can be viewed individually. Later, these layers are made into photomasks that are reproduced on the wafer (see Section 5).

The *logic engineer* looks at the IC from a global viewpoint. He or she uses software programming languages such as VHDL, a hardware description language (HDL, the "V" is just part of the product name), or Verilog to create logic or functional code or logic descriptions. These logic diagrams partition an overall design into functional parts, which can then be used by a circuit designer to create the detailed design for that function. *Synthesized layout* is created by computer aided design tools. A method called standard cell place and route uses computer programs to make the required connections.

Software engineers are needed for chips that must be programmed to communicate with other devices. The most familiar chips of this sort are microprocessors. Their flaws and software bugs are widely reported in the news. Chips used for computer applications are an example of chips that must be programmed after they are manufactured.

Applications engineers put the chips to work in new applications or show how they can be better used in existing applications. Customers often get a lot of help in optimizing the use of the product from these folks.

Technology transfer engineers or a *technology group* will interface with the manufacturing engineers and process engineers who will be producing the chips. The technology group can encompass a wide range of responsibilities from research into new manufacturing processes to reliability standards for existing processes.

The *modeling engineer* is responsible for creating mathematical models and software that mimics the way a certain type of transistor or discrete device will work. They use data that is fed back from the manufacturing environment. Technology engineers as well as by design engineers will use their product, especially if the manufacturing technology is new or unique.

Architectural engineering is the domain of highly experienced engineers who know how to partition and "floor-plan" the "guts" of the semiconductor chip. Large, complicated integrated circuits require many revisions of the initial floor plan. Sometimes an engineer will have to completely change the partitioning in order to fix a problem such as speed (usually too slow) or power (uses too much).

The unsung hero of the design team is the *CAD engineer*. This engineer is responsible for installing and supporting design software for all the types of engineers on the team as well as writing programs to help all the team members do their jobs more efficiently. This team member can also be responsible for evaluating and negotiating the pricing of design and verification software. In small companies, the CAD engineer finishes the integrated circuit design process and performs the final step called *tapeout*.

Cross-functional teams combine several different types of engineers to produce a complete design on a section of the design or of the entire chip.

Functional teams are usually comprised of only one type of engineer. In this case, the new product plan will have a procedure defined that divides the design work into distinct segments that are integrated later.

4.4 The Design Process

Design engineers use software products with names like RTL and Verilog to draw schematics. The idea still comes from the creativity of the engineer. However, today's sophisticated software allows the designer to quickly check his or her work using simulations.

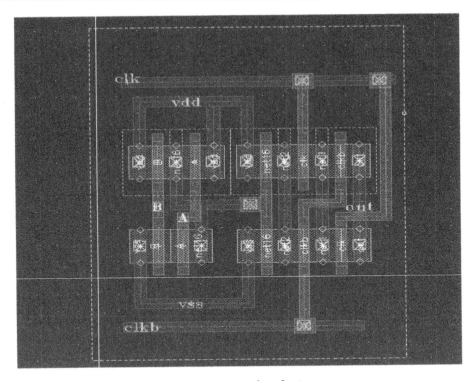

Figure 2-11: Example of a Layout

Simulations are run repeatedly throughout the design process to ensure that the designs will work, that each part of the chip will interface correctly, to check design rules and so forth. Software products like VHDL are used for these simulations.

At the start of the design process, many engineers and other contributors are needed to speed things along. The more people, the better during the early stages of development. Later on, however, fewer people are needed because the tasks are more specialized. Verifications begin to dominate the work and the ability of the layout group to make corrections in the artwork becomes more important, so extra help does more harm than good at the end of the process.

Chips are designed as a whole entity. The schematic is only a symbolic representation of actual electrical components. The layout engineer draws a picture of the actual finished device as seen from the top. Although the chips are manufactured in layers, the design of the chip is not done that way. The chip layout is broken apart into layers by the design software when the design is complete.

4.5 Design Verification and Tapeout

It may sound as if the design process goes relatively smoothly from start to finish. Nothing could be further from the truth. Dozens of things can happen to cause errors that will require changes, corrections and rework of the design. Even simple chips are often designed in blocks that are later wired together. If the electrical parameters do not match exactly, one or all blocks must be reworked. *Simulations* reveal how fast a design will process signals. If the design is discovered to be too slow to process signals, it must be redesigned. Design rules specify such particulars as film thicknesses and conducting line widths. It may be discovered that the design rules were too aggressive and the chip will not be reliable because of manufacturing limitations. The possibilities for bottlenecks are endless.

Design verification is done as often as is necessary to catch all errors and fix them. The new product plan will allow many checks and tests so that errors are found as early as possible in the design process. Verification is done with highly-specialized and expensive software. Originally, verification was done by having the design printed out on paper, and the design engineer and layout designer would use colored pencils to double check the accuracy of connections by drawing over each connection. Today's ICs are too complex for such a simplistic method.

Simulations of how the circuit will operate are run on the finished circuit and during the design of the chip. Simulations are run on small segments of the design, as well as the whole chip design to ensure that each part, no matter how small, will function as intended. Many times the layout will be changed due to a simulation catching a "bug" or error in the design.

Layout Versus Schematic (LVS) is one of the most important tests. It is done to see if the layout artwork matches the schematic and that the connectivity of all the wiring is correct.

Design Rule Checks (DRC) verify that the drawing conforms to the dimensions and object spacing specified in the design rules. This check is done for each portion of the IC, as well as for the finished drawing.

Phase-Shift Mask Check is another test to ensure the file that is sent to the mask-making vendor is drawn for a special type of mask, the phase-shift mask. In particular, poly gate dimensions (see Chapter 5) are now so small that special multiple openings in the photomask are needed to control the phase of the UV light shining through the photomask (see Section 5). This is a new step in the verification process.

Definition: *Tapeout* is the process of turning the layout (chip artwork) into separate layers that will be made into photomasks (see Section 5). The term comes from the procedure used early in the history of the industry when the final step in the design process was to download the layers onto reel-to-reel tapes. The tapes would then be delivered to the mask maker. Today, tapeout essentially consists of sending an encrypted file of the fully verified IC using file transfer protocol (FTP) to the mask vendor.

Photomask and Reticle Preparation

5.1 Introduction

The IC design team has taped out the design of a new product. How is the artwork transferred to the wafer surface? A photolithography process is used for that purpose. Each layer needed to build the chip requires a patterned quartz glass plate: either a photomask or a reticle. These photomasks or reticles are used by photolithographic imaging tools that are a large part of the story found in later chapters.

The production of the photomasks and reticles is discussed in this section. Interestingly, the process most often used to make the reticles and photomasks is an electron-beam lithographic process that is nearly the same as the photolithography used in the fab.

Definition: *Photolithography* is any printing process that uses photographic technology. *Electron-Beam Lithography* simply substitutes electrons for photons. Photolithography is the method used for pattern transfer throughout the IC manufacturing process.

Definition: A *Photomask* is a glass plate with a chrome pattern that can be transferred to the wafer in a single exposure.

5.2 Reticle Substrate Preparation

This account of chipmaking is focused on state-of-the-art processes. In this case, reticles are used exclusively, so there will not be a separate discussion of mask making although the techniques are essentially the same.

Definition: A *Reticle* is a glass plate with a chrome pattern that is used in a step-and-repeat or step-and-scan exposure tool (see Chapter 3). It contains the pattern of only a small portion of the wafer. Multiple exposures are required to transfer the pattern to the entire wafer. These tools have replaced those using photomasks because the image on the reticle can be optically reduced, producing a smaller sized image.

Definition: The Quartz Glass discussed here is fused silica, a high purity form of silicon dioxide (SiO_2).

The reticle is made by patterning a chrome-coated substrate of quartz glass plate. The glass must be defect-free, extremely flat and highly polished. But, more important, it must be transparent to the ultraviolet light wavelengths used to expose the photoresist. All of these requirements are well satisfied by quartz glass.

A thin layer of chromium coats the plate on one side. A glue layer is often needed to ensure that the chrome sticks to the glass and an antireflective coating is needed on top. The total thickness of these films is only about 100 nm.

5.3 Pattern Transfer

Definition: Photoresist (Resist) is a light-sensitive or e-beam sensitive plastic material. Its function is similar to that of the film in a camera, a likeness that will become clear in the next few paragraphs.

The chrome is coated with a thin layer of photoresist. The resist is sensitive to exposure to either an electron beam or a laser. Most ultra large scale integration (ULSI) reticles are produced using an electron beam in a direct-write electron beam system.

Now it is time to utilize the digital code provided by the design group. That code is loaded into the computer that controls the direct-write tool. The layout artwork is reproduced in the resist by the tool, the e-beam is scanned back-and-forth across the resist and the platen on which the glass plate is mounted moves up and down. This intricate control scheme transfers the image into the resist. Exposure to the e-beam or laser energy causes a chemical change in the photoresist.

The pattern is invisible immediately after exposure to the e-beam. A developer solution is needed to dissolve the exposed resist, leaving behind the desired pattern in the resist. Now the chrome is covered with a template that will allow the pattern to be transferred to it.

The next step is to etch the pattern into the chrome. An acid is used that does not attack the resist but easily removes the exposed chrome. A thorough rinse is required after etch, followed by a photoresist strip process to remove the template. The reticle is now complete.

5.4 Inspection and Defect Repair

The reticle must be free from all defects that would degrade its performance before it is shipped to the end user. If there is the smallest imperfection in the pattern, it will be printed on the wafer and every wafer will have many defects, as many as the number of exposures needed to print the entire wafer. This type of defect is called a *repeating defect* and it can devastate manufacturing yield. Particles on the reticle are one type of repeating defect.

Particles are not the only possible defect. Pinholes that are large enough to allow light to leak through can be accidentally etched into the chrome. Luckily, these defects are easily repaired. There are a variety of other types of defects that go beyond the scope of this discussion.

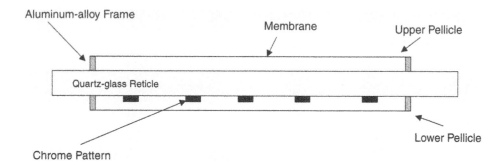

After ensuring that the reticles are defect-free, *pellicles* are installed. The pellicle is a transparent plastic membrane, stretched across a rectangular frame that is larger than the pattern on the reticle. The frame surrounds the chrome pattern and is attached to the glass substrate. The pellicle keeps dust and dirt away from the surface of the reticle. What about dirt on the pellicle? The thickness of the supporting frame holds any particles a distance away from the reticle such that they are out of focus and therefore are not reproduced on the wafer surface.

Now that the pellicles are installed, the reticles can be shipped to the customer and manufacturing of the new ICs can begin.

CHAPTER

3

Forming Wells

Introduction

In this chapter, you will learn:

- What are wells?

- How wells are formed in the wafer

- Glass making: thermal oxidation of silicon

- Photolithography patterning

- Adding dopants to the wafer: ion implantation

Before we commence our discussion of the particulars of making a chip, a clarification is in order. A so-called process flow is a recipe, something like making a sandwich. Our discussion is a generalized CMOS Process Flow that contains all of the required components of the process. A good deal of variation in the order of steps exists, as does the addition or subtraction of some steps; it all depends strongly upon the design of the chip that is being built. So if you go to work and find that your process puts the lettuce on top of the cucumbers rather than the other way around, don't be surprised.

The discussion centers on chips that are used in computers and similar systems. The transistors making up most of the chip, and the chips themselves, have a huge variety of other uses in electrical and electronic circuits that will not be discussed. The computer application is a good example to use because it is easy to understand the transistor's primary function as, quite simply, a switch.

How a Computer Works: Highly Simplified Version

A computer performs billions of operations per second using binary math. Binary (having two parts) math uses only a zero (0) and a one (1); it is called base 2 but as you can see, the figure "2" is not actually used. All letters and numbers are composed of combinations of zeros and ones. The earliest computers used various types of switches to represent the zeros and ones; if the switch was off, it could represent a zero, and then if the switch was turned on, it represented a one. In modern computers the switches are tiny transistors, usually the field effect transistors (FET) that are described in this book. Yes, most of the transistors in a computer are simply switches.

Oh, and what, exactly, are the "billions of operations" being performed each second? It merely means switching the switches on and off! One operation is switching a switch (or group of switches depending on the circuit design) from off to on, or vice-versa.

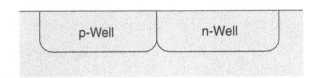

Figure 3-1: Cross-Section of CMOS Well Structure

The complementary metal oxide semiconductor (CMOS) technology uses both n-channel and p-channel transistors on the same chip. The need to produce both types of transistor within the same chip complicates the manufacturing process considerably. It requires that a localized and very thin layer at the surface of the silicon wafer be doped with impurities to adjust the electrical properties of the silicon to match the type of transistor being constructed at that site. A process was developed to produce both n- and p-type doped regions in the appropriate spots, called the *wells*.

Definition: A *well* is a carefully defined area in the surface of the silicon wafer that is doped with either n-type or p-type impurities. The depth of these doped regions may be as little as a fraction of a micron (micrometer) in today's chips. There is a good deal of variation and some chips have wells that are several microns deep.

Figure 3-1 shows the structure that we will discuss in this chapter. The purpose of this process is to create regions of n- and p-type silicon within the wafer surface.

Silicon on Insulator (SOI)

The chipmaking industry is dynamic. It never stops improving itself. This book is a snapshot of the most current technology in use at the present time, but more innovations are constantly being introduced. Silicon on insulator (SOI) is one of them.

SOI is a technique that puts a layer of insulating silicon dioxide very closely under the transistors. This design more effectively isolates each transistor electrically, increasing the chip's speed, reducing leakage currents and improving the chip's resistance to ionizing radiation.

SOI has been used in a small niche of the industry for 30 years, but now it is moving into the mainstream. The reason is that the limits of silicon technology are being reached. Innovation is needed to push existing silicon technology beyond its current performance levels.

The term "well" is a bit misleading. The wells are often much bigger around than they are deep so it doesn't sound like much of a well. However, if you think of it as a region that is deep compared to the transistor structures in its upper portions, then perhaps it makes more sense. The large surface area of the well is needed because often a large number of transistors are made in each well.

The wells are created by adding dopants to the silicon wafer in precisely defined areas. In this chapter, we will discuss the method used to make sure that the dopant only goes where it is needed, as well as the related technologies that help make it all come together.

The n- and p-type wells form the regions where p- and n-channel transistors will be built. Yes, the n-channel transistor is built in the p-well and vice-versa. If that seems confusing, don't worry. As the transistor is built, the function of each of its parts will become very clear.

CMOS Devices

Dynamic random access memory (DRAM) chips need one transistor and one capacitor for each bit of data in memory storage. Many static random access memory (SRAM) chips use six transistors to make a memory bit; obviously, the SRAM functions quite differently than a DRAM. Each has an important role in computer operation.

Another example of a digital circuit element that benefits from CMOS technology is the *inverter*. A CMOS inverter consists of two transistors, one n-channel

---

and one p-channel. In an inverter, a low voltage input is changed to a high voltage output or vice-versa. A high voltage often represents a "1", and a low voltage represents a "0" in logic circuits. Note that "low voltage" means near zero, and "high voltage" means near to the supply voltage in the circuit, which is, in current advanced digital chips, about one volt—not exactly what one would expect to be called a high voltage, is it?

Definition: One *Bit* of data in memory is a zero (0) or a one (1). This term is used in memory chips.

Definition: One *Byte* is the computer term for eight bits. Bytes are combined to make characters or "words" in a programming language.

Terminology Tip: Micron = Micrometer. The metric system is used for measuring dimensions on chips. The micrometer, one millionth of a meter (1×10^{-6} meters) or 1/1000 of a millimeter, is very useful for describing structure sizes on a chip. The practice of shortening the term to "micron" has evolved over the years and is interchangeable with the more formal version.

Since the dimensions have become so small on state-of-the-art parts it is very common today to see the nanometer used for length measurements in addition to the micron. The nanometer is 1×10^{-9} meters (one billionth of a meter or 1/1000 of a micron) in length. Surprisingly, there is no industry slang for "nanometer." Today, the familiar term "nanotechnology" is broadly applied to any technology dealing with microscopic dimensions, not necessarily of the nanometer scale.

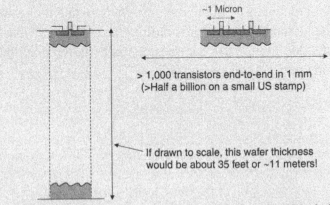

~1 Micron

> 1,000 transistors end-to-end in 1 mm
(>Half a billion on a small US stamp)

If drawn to scale, this wafer thickness would be about 35 feet or ~11 meters!

Roughly 100 transistors end-to-end = Diameter of a 100u hair

Figure 3-2: Relative Size of IC Device Features

A helpful perspective is provided by the diameter of a typical human hair. A thick hair from a person's head is about 100 microns in diameter. A fine hair can be as thin as 50 microns. As you will see, many of the features that are formed on a chip are about one one-thousandth of the diameter of a hair.

Terminology Tip: Product = Wafers. Throughout our discussion, the word "product" will often be used to refer to the wafers moving through the manufacturing line. It is standard practice in manufacturing to refer to whatever it is that is being made as "product." Of course, the chips themselves are the final product so it is OK to use the term for them as well.

Process Flow for Well Formation

1. Grow silicon dioxide on the wafer surface

2. Create a mask or template to define the areas to be doped

3. Add the dopant species by implantation using a particle-beam machine

4. Remove the masking material

5. Heat the wafer to properly distribute the dopant and repair any damage to the silicon crystal lattice

After the wells are created, the wafer is ready for the next step, *shallow trench isolation*, discussed in Chapter 4. Note that some processes form the isolating trenches before creating the wells. If there is any advantage to reversing the order of these two steps, it will be found in the design of the process. At any rate, it does not change either of the steps to any significant degree if they are reversed, so our discussion is not affected.

Initial Oxidation

The beautifully polished, newly-manufactured blank wafers enter the fab. Careful procedures are followed to ensure that minimum contaminants enter the fab with the wafers. Sometimes these wafers are cleaned prior to use in a special wet clean tool, but often the wafers are known to be in an optimum state that is almost free of contamination so the extra step is not necessary. In fact, it is possible that an extra cleaning risks adding contaminants!

Figure 3-3: Silicon Wafers in Carrier (Cassette)

Contamination Levels

It is common practice to use terms such as "contamination free environment." As will be pointed out in this discussion, there are always contaminants and impurities present in trace amounts. One of the greatest challenges in chip-making is to employ designs and manufacturing strategies that minimize the effects of the natural levels of impurities that are always present.

Definition: *Dielectric* is the name of a substance that strongly resists the flow of electrical current; an insulator. Silicon dioxide is a member of this family of materials.

Definition: *Thermal Oxide* is silicon dioxide (SiO_2) produced by the thermal oxidation of silicon.

Purpose: Thermal oxide is a dielectric film (an electrical insulator). It is used in chip structures that insulate conductors, mask implants and etches and performs other protective functions.

Discussion

The material properties of thermal oxide are extremely uniform, chemically consistent and reproducible. This high quality film performs critical functions on the chip and will be discussed later.

A silicon dioxide film called *native oxide* will form naturally on silicon when it is exposed to the air but in that case only a very thin, nonuniform layer forms. The thermal oxidation process produces a very uniform film with the precisely controllable thickness and material properties needed in chip fabrication technology.

The Basic Chemical Reaction:

$$Si + O_2 \rightarrow SiO_2 \quad \text{(silicon and oxygen form silicon dioxide)}$$

What Color is the Wafer?

One of the interesting things about the thin films on the wafer is their color variation. Many of the films deposited on the wafer are transparent. Different thicknesses of films and stacked films cause the wafer to change color throughout the process. Photoresist can produce a beautiful rainbow of color; unfortunately, in this case, it is usually a signal that something has gone wrong.

In the early years of the industry, the machines needed to measure all of the varieties of film thicknesses required for chipmaking had not yet been invented. So the film thickness was checked after oxidation or deposition by holding the wafer up to a color chart hanging on the wall close to the tool. The operator had to be able to compare the wafer color to the examples shown on the chart to determine the film thickness. Some of the texts found in the bibliography contain a table listing the colors of various films based on their thickness.

A Familiar Example

Rust is the product of a chemical reaction. Iron oxidizes or "rusts" in the presence of oxygen and moisture, leaving a coating of iron oxide (Fe_2O_3) on the surface of the iron. Similarly, the process of oxidizing the silicon wafer leaves a coating of SiO_2 covering the wafer.

When charcoal is burned, an oxidation reaction is taking place, too. The fuel is primarily composed of carbon, which reacts with oxygen in the air to form gaseous oxides, carbon monoxide (CO) and carbon dioxide (CO_2). The reaction is exothermic: it gives off heat, allowing the production of essential hamburgers and hot dogs.

It is interesting to note that the ease with which a silicon dioxide film can be formed on silicon, as well as its electrical properties and stability were important factors in the rapid transition from germanium to silicon in the early years of the semiconductor industry. Oxide also has other advantageous properties like the ability to block the implantation of dopants, which will be discussed later.

Silicon dioxide is actually a very familiar substance; it is the principal component of *glass* and several other items mentioned in the next sidebar. Both "oxide" and "glass" are industry slang terms for silicon dioxide.

Another Familiar Example

Sand is perhaps the most familiar example of naturally occurring silicon dioxide. Much of the sand in the world is principally composed of silicon dioxide, some of quite high purity. Another common, naturally occurring form of SiO_2 is quartz, a single crystal form. An important industrial product is "quartz glass," sometimes made by melting high purity quartz crystals but also can be produced today from melting very high purity synthetic silicon dioxide. Quartz glass is widely used in high temperature environments including many applications in the semiconductor industry. The thermal oxide film grown on silicon wafers is chemically similar to quartz glass.

Several gemstones, such as opal and amethyst, are also forms of silicon dioxide. Common bottle and window glasses are about 70% silicon dioxide.

Dielectric materials are the complement to conducting materials that allow the creation of electrical components in the circuit. Several dielectric materials are needed to make chips and each will be explained as it is encountered.

The critical value of oxide is that it is a hard, stable dielectric. Since electrical circuits are being built on the chip, it will certainly be necessary to separate conducting materials from each other with an insulator to avoid an electrical short. Recall that the chip is built in layers, and insulating layers separate conducting layers as required by any electrical circuit. Dielectrics also form the separating medium between the plates of a capacitor and play a critical role in the function of the transistors and other components on the chip.

The electrical properties of oxide are important, but it has other uses too. It may serve as an implant mask, blocking the path of dopant particles that are shot at the wafer in an ion implanter (discussed later in this chapter). Another important use is as an etch mask, protecting portions of a film from attack by etchant chemicals.

What function is the oxide coating performing at this point in the process?

Definition: The *Active Area* is the region on the surface of the wafer where the transistors—the "active devices"—will be built. It must be kept as free as possible from contamination and damage.

Key Point: The initial oxide is a protective coating. Although the film covers the whole chip, the focus of its use is in the "active area."

Definition: *Sacrificial Oxide* is the common name for a thermal oxide that will be removed later in the process. Sacrificial oxides are used as protective layers or occasionally to remove contaminants from the silicon surface. The initial oxidation step produces a sacrificial oxide that is etched off after the wells have been formed.

The initial oxide also influences the amount of dopant entering the wafer from the implanter. The thickness is carefully controlled so that the degree of screening is predictable and uniform.

Another important function of the initial oxide is to prevent "knock-on" of surface contaminants, an effect where the impacting ions from the implanter push unwanted impurities into the silicon crystal where they can move around and damage the transistors.

The Implanter

For clarification, it should be mentioned that the implanter is a particle accelerator that shoots charged particles at the wafer, driving them into the surface. The details of implantation will be discussed in Section 4.

The Oxidation Process

Definition: *Thermal Oxidation of Silicon* is the high temperature reaction of the silicon wafer with oxygen, water vapor, and/or nitrogen oxides. This is usually the first operation to be performed on the wafers in the fab.

Purpose: To form thermal oxide, silicon dioxide produced by direct oxidation of silicon at high temperature.

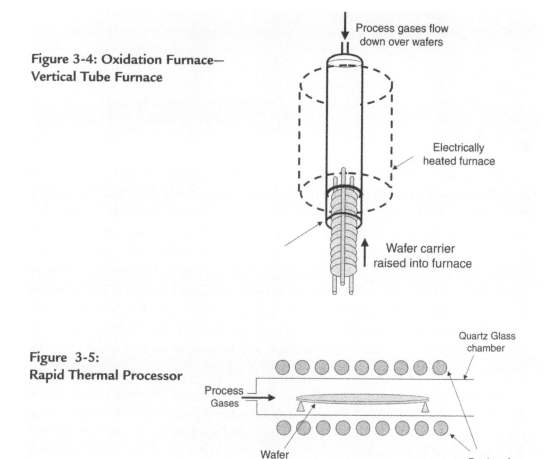

Figure 3-4: Oxidation Furnace— Vertical Tube Furnace

Process gases flow down over wafers

Electrically heated furnace

Wafer carrier raised into furnace

Figure 3-5: Rapid Thermal Processor

Quartz Glass chamber

Process Gases

Wafer

Banks of high-intensity lamps

Discussion

This oxide may have other names, such as sacrificial oxide or pad oxide, and can do several things, depending upon the process. Some of the specifics will be discussed in this and subsequent sections.

Initial Oxidation Process Flow

1. Furnace oxidation

2. Inspection

> A wet clean precedes most film growths or deposition steps. Using various combinations of acids, bases, oxidizers and DI water, surface contaminants are removed.
>
> The temperature is very high during the majority of the oxidation or other film deposition operations; unwanted chemical impurities can move around in the wafer to points where they would ruin many transistors. The phenomenon is called *diffusion*, which is the energetic vibration of the atoms in the crystal as well as the expanded dimensions of the crystal at high temperature create an environment where stray atoms will be moved around in the crystal lattice.

Discussion

Thermal oxidation in a tube furnace is done by placing the wafers in the furnace while it is flushed with gas, often nitrogen, to displace the air. Then oxygen or water vapor is introduced while heating the wafers to, typically, somewhere in the range of 800°C to 1000°C, depending on the process requirements. A furnace oxidation requires that the wafers be placed in special temperature-resistant holders called *boats*, which are slowly inserted into the furnace. Batches of 100 or more wafers are processed at the same time. The furnace is heated by electrical resistance heaters. The slow heating and cooling of the wafers (temperature ramping) is critical to the process to avoid thermal shock which would warp the wafers and damage the crystal structure. Some processing cycles are quite long and often wafers at opposite ends of the furnace display some variation in film characteristics.

Another type of high temperature tool is called a *rapid thermal processor (RTP)*. The RTP is a "single-wafer" system, meaning that it processes only one wafer at a time. The wafer is rapidly heated with high-intensity lamps. The process gases, in this case oxygen or water vapor, are introduced when the wafer reaches the prescribed temperature. After the film is grown, the lamps are extinguished and the process gases are turned off. Water-cooled wafer platens or a flood of nitrogen gas may be used to help rapidly carry heat away from the wafer. A typical RTP system will heat the wafer to 1000°C and cool it back down to room temperature in less than one minute.

The issue of throughput is very important in the manufacturing environment. A large number of wafers must be processed in a given amount of time in order to minimize costs. The competition between batch processors and single-wafer processors has always been fierce, not only because of throughput but also the many other important process specifications. An interesting variety of both of these types of processor, as well as some innovative cross-breeds that combine aspects of both, are found in fabs today.

Thermal Budget

Thermal budget is the term that refers to the amount of heat energy that the wafer will tolerate during the entire process. The devices being built on the wafer will not tolerate high temperatures later in the process. For example, when metal is deposited on the wafer later in the process; that metal will deform or react with its surrounding materials at only a few hundred degrees Celsius. Thermal oxide is grown at about 1000°C. Clearly, it will not be possible to grow an oxide on the wafer after metal has been deposited without ruining the devices. Temperature limitations will be pointed out throughout our discussion.

Inspection

During the inspection phase of the process, measurements are made to see that the film thickness is correct according to the design specifications for the chip. The term "in spec," is industry slang for a properly done processing step. The deposition steps require very close adherence to film thickness and cross-wafer thickness uniformity specifications.

An example of the initial oxide thickness used for some state-of-the art CMOS devices is 15 nanometers, which may also be expressed as 150 Angstroms.

Definition: One *Angstrom* (Å), also called the *Angstrom unit*, is 1×10^{-10} meters in length (one ten-billionth of a meter) or one ten-thousandth of a micron. It may also be expressed as one tenth of a nanometer. Recall that a nanometer is 1×10^{-9} meters or one billionth of a meter.

The Angstrom Unit

The Angstrom unit is named after Anders J. Ångström, a Swedish physicist. At first glance, it seems to be a cumbersome and annoying unit of measure because it is inconsistent with the standardized units of measure, like the micron and nanometer, all of which are spaced three orders of magnitude apart—that is, the next smaller unit is 1/1000th of the larger one (divide the millimeter into 1000 parts and each one is a micron in length).

However, the Angstrom unit has special physical meaning; the radii of all atoms are approximately one or two Angstrom units. According to the Sargent-Welsh Periodic Table, the largest atomic radius is Francium at 2.7Å, the smallest is Neon at 0.51Å. It is especially meaningful for chipmaking since the films deposited on the wafer are extremely thin. This fact is illustrated in the discussion of the Gate Dielectric film found in Chapter 5.

Important Film Characteristics

1. Thickness: The exact oxide film thickness is specified so that the implantation screening effect of the oxide is consistent from wafer-to-wafer and batch-to-batch, while still providing enough protection to the underlying silicon.

2. Uniformity: The wafer will have many individual chips on it and the goal is for each one to operate identically. For that to happen, the cross-wafer film thickness must be kept within tight limits.

As the reader might expect, a large number of product wafers is not committed to a processing tool without double-checking that it is working properly. These tests are performed on a timetable that varies from fab-to-fab, but monitoring the tools is a necessity. Some of the common names for these wafers are *test wafers*, *pilot wafers* or *monitor wafers*. It is not necessary for this discussion to include the procedures involved with these wafers. Suffice to say that they are processed in the standard manner, inspected, measured and if all is well, then large batches of product are committed to the tool.

Photolithography

3.1 Introduction

Definition*: Lithography* is a printing process.

Definition*: Photolithography* is a printing process that uses photographic techniques to transfer a pattern from a photomask or reticle to a coating of photosensitive material, called *photoresist*, on the wafer surface.

Definition*: Photoresist (resist)* is a light-sensitive plastic material that is dissolved in a solvent to form a solution. The resist is dispensed onto the wafer in the form of a liquid, then stabilized (solidified) by a baking step that also affects the photosensitivity and solubility in developer solution as described later in this section. Most photoresist is sensitive to ultraviolet light.

Key Point*: The process of *wafer patterning* is now under way. Before getting into the details of the patterning process, take a look at Figure 3-6.

Figure 3-6: Patterned Wafer

79

To the naked eye, the patterned wafer looks like a bunch of small squares or rectangles. However, when looking under magnification, the real details become visible. The pattern that is transferred to the wafer is found on the *reticle.* Recall that the reticle may have the pattern for one die or many die, depending on their size and the size of the exposure field of the imaging tool (stepper). The terms "die" and "chip" are used interchangeably throughout this discussion. One reticle is required for each layer that needs a mask for implanting or etching. The pattern for most layers appears to be an uncomplicated set of squares, lines and rectangles on the reticle. It is impressive that these simple bits combine to make such a sophisticated device.

Die vs. Dice

The plural of "die" is "dice" (see any dictionary). However, usage in the industry has evolved away from use of the plural form and "die" is commonly used for both singular and plural.

The comparatively wide spaces between each die are called *scribe lines* or *streets.* These spaces must be rather wide because the die are cut apart with a diamond saw. The blade is made as thin as possible, but it is still gigantic compared to the features on the chip (see Chapter 2).

Photolithography Process Flow

1. Coat (Spin)
    a. Surface Preparation
        i. Vacuum dehydration bake
        ii. HMDS (hexamethyldisilazane) application
    b. Spin (resist application)
    c. Soft bake

2. Expose (Step)

3. Develop
    a. Apply developer solution
    b. Post-develop rinse

4. After Develop Inspection (ADI)

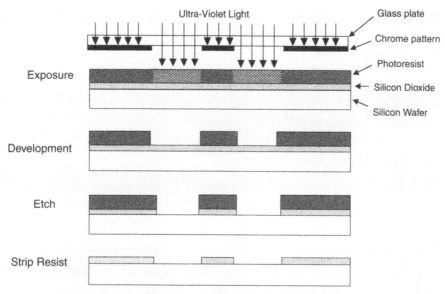

Figure 3-7: The Wafer Patterning Process

The Yellow Room

Now we enter the *yellow room*. Since photoresist is sensitive to ultraviolet (UV) light, the area of the fab where it is processed must be free of UV. Visible or white light that we are all familiar with contains UV at the high frequency end (the blue end) of the spectrum so it cannot be used to illuminate the photolithography processing area. Cutting off the upper end of the spectrum changes the color of the light and makes everything look yellow.

What would happen if white light got into the photo area? No one would know there was a problem until the wafers were put into the developer solution. All of the resist would dissolve in the developer leaving no pattern on the wafer! Why does that happen? Read on.

Definition: An *Implant Mask* is a pattern created in photoresist or other masking material.

Purpose: An implant mask allows the introduction of dopants from an implanter to enter only the specified areas of the wafer. The resist forms a (comparatively) thick layer of plastic that absorbs the speeding ions or neutrals so that they are unable to reach the surface of the silicon.

Dopants are used to create the transistors and some of the electrical components of the chip. The dopants must be placed precisely in the correct areas for the components to work.

3.2 Coat (Spin)

3.2.1 Surface Preparation

When the wafers arrive at photolithography (photo), they enter a low pressure bake oven and are treated to a "sauna" of hot, dry nitrogen gas to drive off any moisture. Then hexamethyldisilazane (HMDS) is dispensed into the chamber in vapor form, where it coats the wafers. The HMDS forms an important "glue layer" that ensures the photoresist will stick to the wafers.

Surface preparation is always a major issue in chipmaking. We would be in real trouble if the layers didn't stick to each other!

3.2.2 Resist Application

The process of coating the top of the wafer with photoresist is sometimes called *spin*, simply because the tool that puts on the resist spins the wafer around. The tool holds the wafer on a spinning vacuum chuck. The resist is dispensed onto the center of the wafer through a tube and is immediately flung off sideways because the wafer is spinning at a rather high speed. Believe it or not, that is the idea. The spin speed is the principle determinant of the thickness of the resist that is left on the wafer; resist viscosity is also an important thickness factor (it is often the thickness of pancake syrup or engine oil). Figure 3-8 shows a spinner.

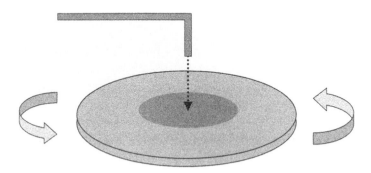

Figure 3-8: Photoresist Coater (Spinner)

3.2.3 Soft Bake

After the resist is applied to the wafer (spin), it goes to a hotplate for a "soft-bake" at about 100°C. Most of the solvent is driven off, leaving a sticky semisolid that serves as a kind of photographic film. Then the wafer moves to the stepper for exposure.

3.3 Exposure (Step)

Definition: *Exposure* is the process of shining ultraviolet light onto the wafer. The wafer is exposed to ultraviolet light just as the film in a camera is exposed to visible light.

The *stepper* is an incredibly sophisticated device. The name "stepper" comes from the way the machine performs its job of exposing the wafer to UV light. A square or rectangular region of a few square centimeters is exposed at a time. The stepper physically moves or "steps" the wafer in preparation for the next exposure.

Watching the machine in action is fun. The wafer is positioned for the first exposure, the shutter opens for a second or less, allowing the UV light to shine on the wafer. Then the wafer is suddenly moved a short distance followed by another flash of light. The process continues, step by step, back and forth across the wafer until the entire surface is exposed. There can be a hundred or more steps needed to expose an entire wafer.

What is truly amazing about the stepper is the extreme precision with which it moves the wafer for each exposure. Remember that each layer required to make a chip must be placed exactly on top of the last layer or everything will be jumbled up and the devices will not work. The tolerance for alignment of the wafer is measured in nanometers for today's devices, requiring the use of lasers to determine the position of the vacuum chuck assembly that holds the wafer.

In more familiar terms, the stepper positions the wafer with a tolerance of closer than 1/1000 of the thickness of a hair. That's precision!

The stepper is essentially taking a picture of the reticle. The film in which the picture will be developed is the photoresist on top of the wafer.

Figure 3-9: Reticle Projected onto Wafer

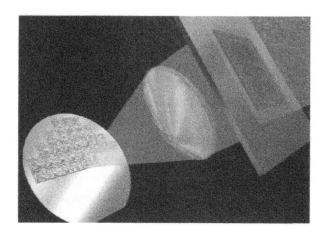

Definition: The *Exposure Field* or simply, the field, is the area on the wafer surface that is exposed to UV light at each exposure. The majority of steppers in today's state-of-the-art fabs use a scanning technique that sweeps the light quickly across the field. These tools are called "scanners" and use a step-and-scan technique.

The term "flash" is another slang term for the exposure field that applies to steppers that expose the entire area of interest as if a camera shutter opened and closed. This stepper technology dominated the industry for many years and is still in use today.

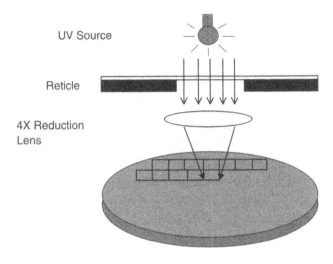

Figure 3-10: Stepper Prints Image in Resist

3.4 Develop

Definition: *Deionized Water (DI Water)* is water that has been processed to remove impurities, especially those in an ionized state, such as sodium (Na^+) and calcium (Ca^{++}). When water is needed, only DI water is pure enough to be used in the chip manufacturing process.

Developing the pattern in the resist is very much like developing the film from a camera. Exposure to UV light chemically changes the resist. Liquid developer solution is dispensed on the resist; it dissolves away only the areas of resist that were exposed to UV light. Then a thorough rinse with DI water removes any leftover chemicals and leaves a precise pattern behind. Development takes place in another tool that spins the wafer on a vacuum chuck to improve the consistency of chemical distribution.

In this case, the remaining resist is the protective mask that blocks the implant. Only the open areas in the resist will receive dopant.

3.5 After Develop Inspect (ADI)

The entire manufacturing process is interspersed with inspections of the wafers to check critical dimensions and to catch defects before they ruin the product. This is the only time that inspections will be discussed unless there is some special insight to be gained into the chip or manufacturing process.

The *after develop inspection (ADI)* includes a visual inspection of a sample of wafers under the microscope, measurements of the feature sizes created in the resist, scanning electron microscope (SEM) inspections, measurements and checks for particulate contamination. There are even SEM cross-sections (done infrequently) that require breaking a wafer. Most of the inspections and measurements are done by automated tools that can recognize the pattern in the wafer; go right to the programmed spot for the inspection; compare patterns to find defects and contamination and many other jobs that formerly required an operator to handle the wafers (see Chapters 2 and 10).

An example of a completed photoresist pattern is shown in Figure 3-11.

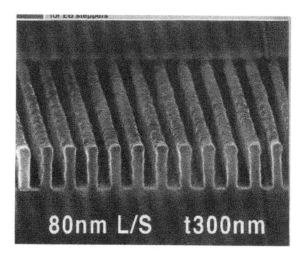

Figure 3-11: Scanning Electron Micrograph (SEM) of Developed Photoresist Image (Nikon)

Ion Implantation

Implant Process Flow

1. Mask for n-type implant

2. Implant n-type dopants

3. Remove mask

4. Well Diffusion (sometimes done only once at the end)

5. Remask for p-type implant

6. Implant p-type dopants

7. Remove mask

8. Well Diffusion

The process flow for Well Implant shows that it is a two-part operation. Since CMOS parts combine n-type and p-type devices as explained earlier, two different implants and masks must be used (see Figure 3-12).

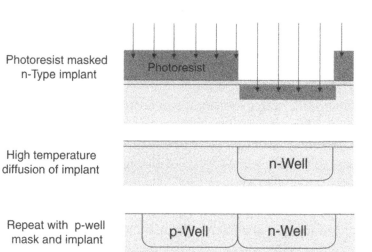

Photoresist masked n-Type implant

High temperature diffusion of implant

Repeat with p-well mask and implant

Figure 3-12: Sequence of Steps for Well Doping

Implant Overview

The implanting process is very straightforward. It simply involves shooting ions at the wafer so that they embed

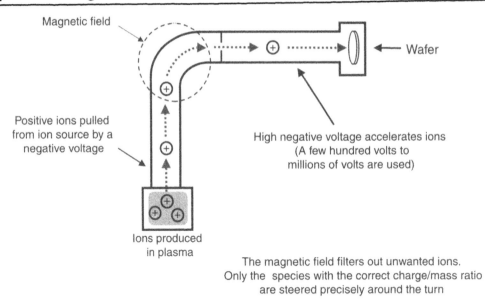

Magnetic field

Wafer

Positive ions pulled from ion source by a negative voltage

High negative voltage accelerates ions
(A few hundred volts to
millions of volts are used)

Ions produced
in plasma

The magnetic field filters out unwanted ions.
Only the species with the correct charge/mass ratio
are steered precisely around the turn

Figure 3-13: Schematic of Ion Implanter

themselves in it. The technology allows very precise amounts of dopant to be added to the wafer as well as a good deal of control of the depth that the ions or atoms travel into the wafer.

Generally, charge accumulation on the wafer surface is unacceptable since it may damage the wafer so neutral atoms are implanted rather than those with a charge on them. The ions pass through a region at the end of their run through the implanter where there are many electrons available. The positive ions grab an electron, replacing their missing one (opposite charges attract), and become a neutral atom. The atom's momentum carries it forward and into the wafer surface.

The implanted species travel a very short average distance into the silicon, usually only a fraction of a micron. In addition to controlling the areas where the dopant atoms enter the wafer surface with the photoresist patterns, the depth of penetration of the atoms must be controlled. The acceleration voltage used in the implanter controls this depth, which also depends on the size of the atom being implanted. For the same depth, a higher acceleration voltage must be chosen for a large atom than for a small one.

Channeling

Since the crystal planes of the silicon are aligned with the surface of the wafer, the wafer is angled 7–10° to avoid channeling, where the accelerated ion or atom follows a major crystallographic axis, slipping between atoms of Si and moving deep into the wafer. Control of the depth of implant relies partly upon the implanted species colliding with silicon atoms.

By way of comparison, orchards bear a strong resemblance to crystals. Driving down the road next to an orchard is a good way of imagining what the implanted atoms might see when approaching the silicon wafer. If coming from an angle that is off-axis, the trees look like a jumble of random arrangement. However, when the spaces between the rows of trees appear you can see clear across the orchard. The crystal is like the orchard and the spaces between rows of atoms are the major crystallographic axes. It is easy to see why channeling atoms travel so far.

Uniform coverage of the wafer surface is another key concern. Different implanter manufacturers control dopant distribution in a variety of ways. Figure 3-14 shows one example, a tool that holds a number of wafers around the circumference of a large cartwheel-shaped assembly. During processing, the cartwheel rotates at a high speed, passing the wafers in front of an opening that exposes a very small part of the wafer at a time to the implanter beam. This motion provides vertical scanning; the beam is scanned horizontally. The entire wafer is implanted, a little at a time so the dose can be carefully metered and the temperature can be kept under control. High dose, high energy implants can overheat the wafer and damage the resist mask.

Figure 3-14: Ion Implanter Platen (Applied Materials)

Crystal Damage Repair and Dopant Diffusion

As might be imagined, the bombardment of the wafer damages the silicon crystal lattice. The silicon atoms are knocked out of place and the orderly bonds are broken. In high energy, high dose implants, the crystal structure at the surface of the wafer is completely lost; it is said to become "amorphized." In addition, there are now many dopant atoms added to the mix.

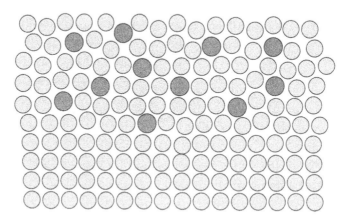

Figure 3-15: Silicon Crystal with Dopant Atoms as Implanted

The desired result of the doping operation is to replace some of the silicon atoms with dopant atoms (see Chapter 1) while maintaining the uniform crystal structure. The chip will not work if the crystal is not repaired with those substitutions included. Luckily, the crystal can be reassembled to match the crystal structure of the bulk of the silicon wafer by heating it to a temperature that is well below the melting point of silicon; the melting point of silicon is 1414°C—that's hot! The lattice can be restored at a temperature of 800°C to 900°C in a short time. This heating operation is called *annealing*.

The annealing operation repairs the damaged crystal. It also causes the dopant atoms to move a known distance through the silicon. This natural movement, or diffusion, results in distribution of the dopant through a wider area than is allowed by the implant alone. The vibration of the atoms in the solid at a given temperature moves the dopant slowly through the silicon in a very controlled manner, permitting the exact dimensions of the conductive region to be defined.

When the dopant atoms are in place and the crystal structure is restored, the electrical properties of the silicon have been altered. At that point, the dopant is said to be "activated." More formally, "activation in the host lattice" has been accomplished as a result of the annealing. Now the regions of implanted silicon are more conductive than the starting wafer and it is either n-type or p-type depending upon the dopant used.

In the particular case of the well process, higher temperatures are used and applied for a fairly long time because the dopant atoms must diffuse relatively long distances to form the well. This step is often called a *well diffusion* or a *well drive-in* rather than an anneal. In later steps, diffusion distances must be kept short and low temperature annealing, as discussed above, is needed.

If the implant is shallow, with the highest concentration of atoms near the surface of the wafer, the anneal will move the atoms downward. That type of well will have a high concentration of dopant at the surface and a decreasing concentration of dopant moving down into the wafer.

A second doping strategy uses a deep implant. The implanted atoms are driven deeply into the wafer, leaving a region above that is essentially undoped. In this case, the annealing moves many of the dopant atoms upward, toward the wafer surface. This forms a "retrograde" well in which the concentration of dopant increases moving down from the surface, the opposite of the shallow implant well. There is an advantage offered by the retrograde well: less chance of "latch-up," a condition that may occur during operation that causes a local high current flow that may damage or destroy the chip.

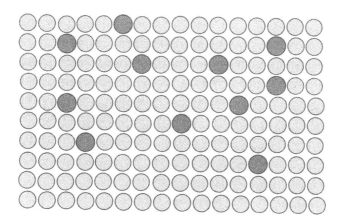

Note: Schematic representation only. Ratio of dopant atoms to silicon atoms actually much smaller than shown

Figure 3-16: Dopant Distribution after Anneal

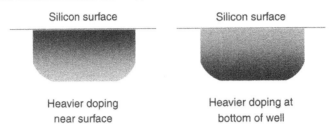

Silicon surface   Silicon surface

Heavier doping   Heavier doping at
near surface   bottom of well

Figure 3-17: Doping Profiles in Wells

Process Notes

The dopant species most commonly used to create n-type wells is phosphorous. The major p-type dopant is boron.

It is not surprising that some implants involve a large amount of energy transfer to the wafer, which raises the wafer temperature. The principle issue with wafer heating is damage to the photoresist mask. If resist is exposed to temperatures above levels specified by the manufacturer it suffers progressive damage; it will melt or "flow," as the temperature rises, then harden or "reticulate" and eventually, at a high enough temperature, will burn. There is also some damage to the chemical structure of the resist because of the bombardment from the implanter. Damaged resist is hard to remove from the wafer; if conditions become too extreme, the mask will become distorted and deformed and the parts will be ruined. Minimizing resist damage is an important process control issue.

Although well implant does not generally produce enough heating to damage the resist, later implantation steps such as the source/drain implant discussed in the next chapter require special procedures to reduce the risk to the resist mask.

It is often necessary to use both a plasma etch operation called *ashing*, and a wet chemical stripping solution after implant to remove the resist.

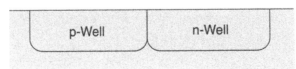

p-Well   n-Well

Figure 3-18: Cross-Section of CMOS Well Structure

Now that the wells have been formed, the wafers are ready to move to the next operation, shallow trench isolation (STI). The trenches will electrically isolate every transistor from its neighbors so that no electrical interactions occur that would ruin the circuits.

CHAPTER

4

Isolate Active Areas (Shallow Trench Isolation)

Introduction to
Shallow Trench Isolation

In this chapter you will learn:

- How transistors are electrically isolated in the chip
- How silicon nitride is deposited
- Some of the details of plasma etching
- Some of the details of chemical mechanical polishing

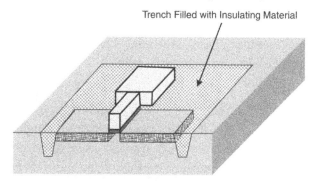

Figure 4-1: Shallow Trench Isolation

Figure 4-2: Shallow Trench Isolation Cross Section

Definition: A *Shallow Trench* is a sort of moat that is dug around every transistor on the chip. The dimensions of the trenches will vary somewhat based upon the design of the chip.

Why are the trenches called *shallow*? Compared to the depth that the transistor goes into the wafer, they are pretty deep, as will be illustrated shortly. The reason for the terminology is that there are trench structures sometimes used for other purposes, such as capacitors for DRAM's, that are quite a lot deeper. So the term simply reflects a generalized comparison of similar chip structures.

Purpose: *Shallow Trench Isolation (STI)* is a technique for electrically isolating each transistor. Trenches are formed around each transistor on the chip. The trenches are filled with silicon dioxide, an electrical insulator, which prevents most unwanted electrical current from flowing between adjacent transistors.

Discussion

Why do adjacent transistors need to be electrically isolated? Because they are electrical devices and will communicate with each other in unwanted ways like shorts, latch-up and cross-talk. These problems will ruin the chip. Some examples will be discussed later in the chapter.

Can't the transistors just be kept separated? Recall from Chapters 2 & 3 that the silicon wafer, which is part of the chip, is doped to make it conductive. Since the transistors are built mostly down into the wafer, they would need to be separated by a rather long distance to reduce the current flow between them. But, the transistors must be packed tightly together in order to fit many millions of them on a single chip, so wide separation of the transistors is not an option.

How good is the isolation provided by a trench? Shouldn't the transistor be completely surrounded by insulation like any other electrical device? In other words, won't the electricity flow under the trench?

The insulation does not have to go underneath the transistors. The isolation under the transistors is taken care of by the presence of p-n junctions. When the chip is operating normally, those p-n junctions act like insulating layers. Review p-n junctions in Appendix A to see how current may flow through these junctions and how it is inhibited from flowing.

However, as you might suspect, there is a very weak connection between the adjacent transistors that is felt underneath the trench. There are mechanisms that can result in unwanted current flow which can cause faulty operation or even chip destruction. Proper chip design and construction minimizes that risk.

Shallow Trench Isolation Process Flow

1. Grow Pad Oxide

2. Deposit Silicon Nitride

3. Photolithography Patterning

4. Form the Hard Mask: Transfer Pattern to Dielectric Films with Plasma Etch
   a. Silicon Nitride Etch
   b. Pad Oxide Etch

5. Etch Trench in Silicon Substrate

6. Fill Trench with Silicon Dioxide
   a. Grow Thermal Oxide Liner
   b. Deposit CVD Oxide Fill

7. Remove excess oxide with chemical mechanical polishing (CMP)

8. Remove remaining nitride and oxide with wet chemical etch

Discussion

Shallow trench isolation (STI) is a very comprehensive process. It includes a majority of the operations found in the fab: thermal oxidation, thin film deposition, photolithography, plasma etch and CMP. STI is especially interesting because in the process of building the trench structures, the interrelationship of these fab procedures is clearly illustrated.

The special requirements of STI are outlined below. Each element of the Key Points will be explained in detail in later sections.

Key Points in the STI Process

1. Chemical mechanical polishing (CMP) must be used to remove excess oxide after trench fill, leaving a nearly flat surface on the wafer.

2. Silicon trench etch uses an inorganic sidewall protective coating called *passivation* to control the shape of the trench. The more common organic passivation could leave contamination behind, so no carbon source is used at that stage.

3. A hard mask for the trench etch must be used in place of the photoresist mask to eliminate organics and provide good trench edge definition. The dielectrics, silicon nitride and pad oxide, are good hard masking materials.

4. The resist is used as the mask for the hard mask etch, but must be removed before the trenches are etched.

5. Anti-reflective coatings are included in the process to aid photolithography but are not illustrated in the figures.

Pad Oxide Growth

Definition: *Pad Oxide* is thermal oxide grown on top of the silicon wafer. It is typically about 150 Angstroms thick. It forms a protective pad between the next layer, silicon nitride, and the underlying silicon wafer surface.

Purpose: The pad oxide prevents damage to the silicon surface that would happen if silicon nitride were deposited directly on the silicon.

Discussion

Silicon nitride expands at a different rate than the silicon wafer. The pad oxide allows the nitride to float above the silicon so that the nitride applies little mechanical stress to the silicon surface during the thermal cycling needed for the process. The area underneath the silicon nitride is where the transistors are to be built as mentioned in Chapter 3; it is the *Active Area* where the *Active Devices* (transistors) are located. The silicon crystal structure is important to the proper functioning of the transistor and must be kept in good condition.

The pad oxide is thermal oxide just like the sacrificial oxide discussed in Chapter 3. In fact, the thickness is about the same in both cases, about 150 Angstroms.

Figure 4-3: Pad Oxide Cross Section

Please feel free to refresh your memory by returning to Chapter 3 to read the discussion of thermal oxidation.

Silicon Nitride Deposition

Definition: Silicon Nitride (Nitride) is a hard dielectric film.

Purpose: In this application the silicon nitride is used as an etch mask to block the etching chemicals from attacking masked-off areas on the wafer surface. This is the job of a "hard mask" as discussed in the next section. It also helps to protect the active areas from physical damage that might be caused by the CMP tool, as discussed in Section 8, although that is of secondary importance to the primary function of the film.

Why not just use photoresist as the etch mask? That question is answered in some detail in Section 5, Hard Mask Formation.

Definition: Chemical Vapor Deposition (CVD) is the process of depositing a thin film on the wafer by exposing it to gaseous chemicals (said to be in the "vapor phase"), which react with one another to form the desired layer material.

Definition: Low Pressure CVD (LPCVD) deposits a film in a low pressure environment, typically 0.25 to 2.0 Torr (around one-thousandth of atmospheric pressure) at temperatures ranging from 400 to 850°C. The LPCVD reactor relies solely on thermal energy to initiate the chemical reaction that results in deposition. Units of pressure are discussed later in this chapter and in Appendix A, Science Overview.

Definition: Plasma Enhanced CVD (PECVD) is deposition carried out in a plasma reactor. The energy of the plasma is used to assist the chemical reaction, which will then proceed at a lower temperature than in LPCVD. Wafer temperatures in the range of 100–350°C are used for PECVD.

Chemical Reaction

$$SiH_2Cl_2(gas) + NH_3(gas) \rightarrow Si_3N_4(solid) + NH_4Cl(gas) + H_2(gas)$$

The above chemical reaction involves the most commonly used chemicals for the LP-CVD silicon nitride deposition process. Dichlorosilane reacts with ammonia to give silicon nitride, ammonium chloride and hydrogen. Ammonium chloride is a solid at room temperature, but is a vapor at the temperature of the reactor.

Discussion

This nitride film will be about 1,500 Angstroms thick.

Silicon nitride is like a number of the materials used in chipmaking in that its chemical composition can vary but it will still function effectively. The ratio of silicon to nitrogen varies and it can contain a substantial amount of hydrogen depending on the way it is deposited. The silicon nitride film for the STI process is deposited without using a plasma, with the wafers at a high temperature of about 750°C and has a composition close to the theoretical formula of Si_3N_4.

In other applications later in the wafer fabrication cycle, a temperature of 750°C would damage the devices, so plasma-enhanced CVD (PECVD) is used to provide a lower temperature process, usually in the neighborhood of 400°C or below.

Many CVD processes depend simply on heating the wafers to a high enough temperature for the deposition reactions between the gases to occur on the wafer surface, but in cases where device damage is an issue, another approach is needed. Using the electrical energy of a plasma can help the reactions along so that deposition will proceed at lower wafer temperatures.

A plasma nitride may have a hydrogen content of 20% or more and will have properties that differ from those of the high temperature material. For example, it will have different optical properties and dissolve much more rapidly in hydrofluoric acid solutions.

In most CVD applications, the film covers all kinds of structural features already formed on the wafer. An important factor in all thin film deposition processes is how well the film conforms to the wafer surface. In particular, it can be difficult to fill small holes and very narrow trenches. Also, films deposited by various techniques are conformal to varying degrees. Films deposited by physical vapor deposition (sputtering) have different conformality characteristics from those of CVD films, and the conformality of CVD films varies with the chemistry used.

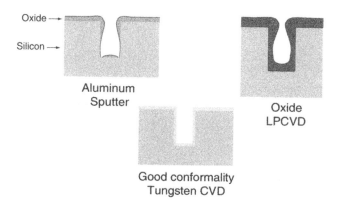

Figure 4-4: CVD Conformality and Fill Characteristics

Silicon nitride has a number of uses and it will appear often in the chipmaking process.

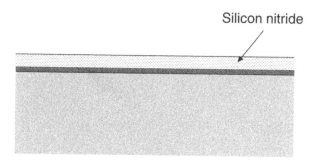

Figure 4-5: Nitride on Pad Oxide Cross Section

Photolithography for Photo/Etch

This section begins the discussion of a new application for photo/etch—the formation of a hard mask. Some photolithography points are reiterated in this section from Chapter 3 to help keep the discussion in context.

Definition: Photolithography is the process of transferring a pattern from the reticle to the photoresist, which will then act as a mask or template. This time the pattern in the resist will be etched into the underlying film, completing the pattern transfer.

Definition: Photoresist (resist) is a plastic material that is sensitive to ultraviolet (UV) light.

Purpose: The photoresist acts as a template, a protective mask that will only allow the underlying film to be etched away from areas not covered with resist. When used this way, it is called an *etch mask*. Recall that the photolithography process in Chapter 3 produced patterned resist that served as an implant mask.

In STI, the resist is used to pattern the nitride and oxide layers. This is an interesting combination process that is used to produce something called a *hard mask*, which is discussed in the next section.

Photolithography Process Flow

The process flow was covered in Chapter 3. There are no changes in that discussion for this application. Here it is again for reference.

1. Coat (Spin)
   a. Surface Preparation
      i. Vacuum dehydration bake
      ii. HMDS (hexamethyldisilazane) application
   b. Spin (resist application)
   c. Soft bake

2. Expose (Step)

3. Develop
   a. Apply developer solution
   b. Post-develop rinse

4. After Develop Inspection (ADI)

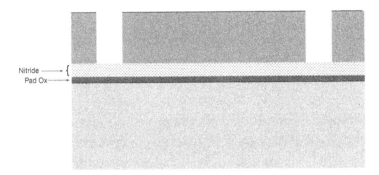

Figure 4-6: Resist Pattern for Trench Etch

Hard Mask Formation Using Plasma Etch

5.1 Hard Mask Overview

Definition: A *Hard Mask* is a mask most commonly made out of either silicon nitride, silicon dioxide, or both.

Purpose of Hard Mask: A hard mask is often used as an etch mask and sometimes as an implant mask. In this case, the mask protects the underlying surface from attack by the etchant chemicals, and only exposed surfaces are etched away. The hard mask substitutes for photoresist in instances where the resist is unable to do the job required by the process.

Discussion

Why is the resist unable to do the masking job for etching the silicon trenches? Most of the time the photoresist is perfectly adequate to use as an etch or implant mask—in fact, it is used as the etch mask to etch the hard mask. But in a few cases there are requirements in the process that make it inadvisable to use resist. Trench etch is one of those cases.

The problem with using photoresist as the etch mask for trench etch is two-fold:

1. The resist will be etched away before the trenches are completed

2. The trenches will be coated with organic (carbon-based) chemical deposits that could contaminate the devices

The chemistry of trench etch will be discussed in more detail later but here is an overview of the major issues.

Briefly, chlorine is the principle reactant used to etch the silicon trenches. The chlorine attacks the resist mask and etches it away rather rapidly. For most applications, a thick enough resist coating is used so that the whole mask is not destroyed before the feature is formed. But in this case, the trenches are quite large compared to most features on the chip; a long etch time is needed to form the trench because such a large amount of silicon must be removed. The chlorine attack is aggressive enough so that all the resist would be etched off long before the trenches were finished. Of course, when the template (mask) that is to be reproduced on the wafer is destroyed, the feature is also ruined. So a tougher masking material must be used.

It should not be overlooked that although the resist is attacked by the etch chemicals, some of the carbon-containing products are re-deposited on the wafer. The amount of re-deposition is quite small, but it is enough to leave significant amounts of trace chemicals that pose a potential contamination hazard that will be discussed later in this section.

Silicon nitride is a very tough material and stands up well to the chlorine chemistry. The pad oxide underneath the nitride is also a tough dielectric and likewise etches very slowly in chlorine chemistry, although its function is primarily to separate the nitride and silicon as discussed earlier.

Of course, the hard mask must be formed in the nitride and oxide films using plasma etch, so a mask is necessary for that part of the operation. Photoresist works very well for this masking operation. Any remaining resist is removed before progressing to the trench etch.

There is another issue regarding chemical contamination. Nitride and oxide etch recipes usually have a chemical component containing carbon, and when combined with some minor resist erosion (recall that resist is mostly carbon), forms an organic protective layer called *passivation*. Discussed later in this chapter, these deposits are important to help control the etch and allowing the production of vertical sidewall profiles so critical to the device.

There is some concern that if any residual organic passivation is left on the sidewalls of the trenches, that later in the process, it could overheat enough to harden or burn. That residue could not be removed in the succeeding wafer cleaning steps and so becomes a contaminant that could endanger the functioning of the device. Therefore, all of the resist and most of the organic deposits are removed from the wafer prior to etching the trenches.

How then are the trench sidewalls passivated? Read on!

5.2 Plasma Etch Overview

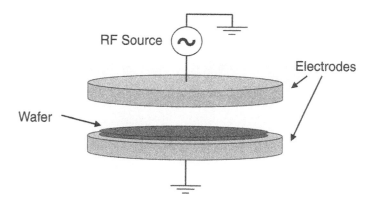

Figure 4-7: Typical Plasma Etch Reactor

Definition: Plasma Etch is the use of a plasma, generated with suitable gases, to chemically remove material from the wafer surface.

Purpose of Plasma Etch: The purpose of plasma etch is to reproduce the etch mask in the film(s) on the surface of the wafer or to chemically remove material from the wafer surface in a less specific way. The mask pattern may be transferred to a film or stack of films that have been deposited on the wafer or etched into the silicon wafer itself. For many critical manufacturing steps, the etch process must also be able to shape the feature being produced on the wafer surface in a specified way.

Plasma etch is also used to chemically remove films that completely cover the wafer. Most of our discussion will center on the most critical applications of plasma etch, which results in properly shaped features produced on the wafer.

The plasma etch process includes both the chemical removal of material and the deposition of a thin coating of material to protect selected areas and assist in creating the desired features on the wafer. Physical bombardment of the wafer with accelerated ions is often utilized to assist in the process although there are several plasma etch operations that do not require bombardment.

Since plasma etch uses gases rather than wet chemicals, it is often referred to as "dry etching."

Definition: Wet Etch is the use of liquid etchants to remove material from the wafer surface. Wet etchants produce an isotropic sidewall profile.

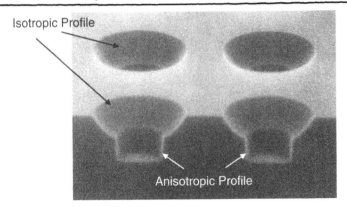

Figure 4-8: Isotropic and Anisotropic Etch Profiles

Historical Perspective on Etch

For about the first twenty years of the semiconductor industry (which started around 1950), essentially all etch processes used in device fabrication were wet, using a wide variety of chemical solutions. Then plasma etches were developed, first to etch materials that were difficult to wet etch and later to etch very small device features that could not be formed properly with wet etching. In current technology, almost all the etches used are plasma processes.

As you will see in a moment, the topic of plasma etch is quite broad. To get a good picture of the entire process requires a discussion of several different technologies, some interesting physics and chemistry and a little history of why and how the technology developed. More information can be found in Appendix B, Plasma Etch Supplement to Chapter 4, but do not hesitate to do some additional reading on plasma etch. Several good books are referenced in the Bibliography.

Definition: *Plasma* is an electrically charged gas. The atoms and/or molecules of gas are made into ions (see Appendix A) by the processing tool.

The etch reactor produces about 1% or less of charged particles mixed with a neutral gas—just right for the process requirements. The plasma produced by processing tools is referred to as a "partially-ionized gas." This distinction is made because, technically speaking, a plasma is made only of ionized gas particles. It is much easier to use a more relaxed definition when discussing plasma processing.

Cascade Effect

The cascade effect is the term for the avalanche of collisions between electrons in a plasma reactor. First, one stray electron collides with another that is bound to a molecule, knocking it off and thereby producing an ion. Now there are two electrons being accelerated by the reactor toward other molecules. When they both knock off another electron, four free electrons exist—then eight, then sixteen—you get the picture.

Definition: Passivation is a chemical coating produced in the etch reactor that is deposited on the wafer. Many plasma etch recipes include a chemical combination that produces this material. The coating "passivates" or protects some wafer surfaces from attack by the etchant chemicals and can offer some protection from bombardment too.

Sometimes the term "sidewall passivation" is used since the protection of the feature's sidewalls from chemical attack is so important to forming the electrical devices on the chip. It is important to remember that the passivation does much more than that, as will be discussed later in this section.

Definition: Photoresist Stripping or "Ashing" is the removal of photoresist from the wafer surface using a plasma. Ashing is actually the etching of photoresist, unlike liquid (wet) resist strip, which is done in a bath of liquid chemicals. The principal gas used for this etch is oxygen, which is used to oxidize or "burn" the resist off of the wafer, hence the name. However, it is somewhat misleading because no actual ash is formed; the reaction products are gases and so can be removed by the vacuum system.

Other trace gases may be added to enhance the process.

Purpose of Photoresist Ashing: Resist ashing is needed to remove the remaining resist mask after a processing operation like plasma etch or any other situation requiring the removal of resist from the wafer. The oxygen plasma is good at removing any other organics that were deposited on the wafer during processing, an important wafer cleaning issue.

Discussion

The use of plasma etch is required for making today's chips. Here are some of the advantages of plasma etch:

1. Plasmas make highly reactive chemical species

2. Anisotropic (directional) etch capability—vertical sidewalls are required

3. Ion bombardment is available to speed up chemical reactions and help to shape features

4. Can deposit a surface passivation film that enhances several aspects of the etch

5. Formation of tiny features—liquids may not enter microscopic openings and therefore are not reliable for producing the smallest features

6. Uses a much lower quantity of chemicals than wet etch and the effluent, or waste chemical, is easier to deal with (more environmentally friendly)

Process Flow for STI Plasma Etch

1. Etch silicon nitride

2. Etch pad oxide

3. Remove remaining resist using an oxygen plasma

4. Etch trench in silicon (discussed in the next section)

5. Wet clean to remove inorganic passivation

Etching Glass

Anyone taking an art class on etching glass will be very familiar with the basics of this process. In the art class, the plate of glass to be etched is coated with wax. Then the artist scratches off the wax in the spots where it is desired to etch the glass. An acid is poured onto the glass, and after a specified time the acid is rinsed off. Then the wax can be removed—and voila! A beautiful work of art has been created.

The first chips to come out in the 1950s were made in somewhat the same way. Photoresist was previously used only in the PC board industry and to some extent in making discrete transistors. However, its use was adopted early on in IC manufacturing and applied to the wet etching processes used at that time. Plasma processing was introduced later, and photolithography grew ever more sophisticated.

Chemical Reactions

The ease with which a chemical reaction takes place depends a lot upon the strength of the bonds holding the reactants together. For example, the strength of the chemical bonds holding the silicon to the oxygen in silicon dioxide is more than double the strength of the bond holding the silicon atoms to each other in single crystal silicon. Needless to say, silicon dioxide etches much more slowly than silicon in a given set of conditions. In this section, the methods used to increase the etch rate of silicon dioxide are discussed.

Hard Mask Etch Strategy

Here is a condensed summary of the strategy employed to etch dielectrics that are discussed in this section:

1. Use highly reactive chemicals

2. Add energy through ion bombardment to increase rate of reaction

3. Passivate with organics to aid selectivity and to help control feature shape

4. Sputter off organic passivation from horizontal surfaces with accelerated ions to allow reactant access to reaction sites on the material below

$$SiO_2 + CF_4 + CH_2F_2 \rightarrow SiF_4 + CO$$

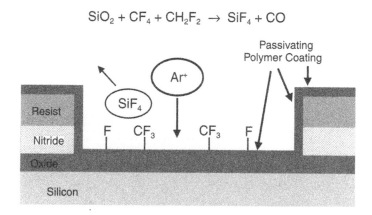

Figure 4-9: Chemical Soup in a Plasma

Figure 4-9 is an attempt to summarize all of the major activities going on at the wafer surface. More is added to this discussion at the end of the section. The large positive argon ions are shown bombarding the surface, adsorbed chemicals are ready to react when enough energy is added by the bombardment. The organic passivant is sputtered off by the argon ions. The volatile products of reaction are shown leaving the surface and being pumped out of the reactor. The dynamic balance achieved in plasma etch is truly impressive.

5.3 Etch Chemistry: Silicon Dioxide and Silicon Nitride

It has already been pointed out that the strong bonds holding oxide together are hard to break. The same goes for nitride. The low reactivity of the films means they will have a slow etch rate. A highly reactive etchant is required to boost rate and increase throughput.

Oxide and nitride are etched with fluorine (F) chemistry. Fluorine is part of Group VII on the Periodic Table (right next to the Noble Gases). The elements in that group are highly reactive. In fact, fluorine, at the top of the list, is one of the most reactive substances in the universe. It is that property that makes fluorine so useful in etching these two silicon compounds. Fluorine will produce the fastest etch rate possible.

A Typical Oxide Etch Chemical Reaction

Plasma etch reactions are often quite complicated with a number of steps involved. Following is a simplified example for oxide etch:

$$SiO_2 + CF_4 \rightarrow SiF_4(g) + CO_2(g) + CO(g)$$

Key Point: The plasma etch chemical reaction must have volatile products or the reactant will be blocked from the reaction sites by a layer of solid material, stopping the etch reaction. For example, a fluorine plasma does not work for etching aluminum because, unlike silicon tetrafluoride, aluminum fluoride is not volatile.

In the plasma:

	Radical	Ion	Excited State
$CF_4 \longrightarrow$	CF_3	CF_3^+	CF_3^*
	CF_2	CF_2^+	CF_2^*
	CF	CF^+	CF^*
	C	C^+	C^*
	F	F^+	F^*

* indicates Excited State

Figure 4-10: CF_4 Breakdown in the Plasma

Many chemicals may be combined to etch silicon compounds. These complicated etch recipes have been developed to produce difficult feature shapes. A very important factor is the control of reaction products that coat the wafer and act as passivants. These compounds help to control the shape of the features being created on the wafer as well as enhancing the selectivity to underlying layers.

Definition: A *Radical* is a group of two or more atoms that tend to enter into chemical combinations as a unit. The *Free Radical* is a form of radical that has one or more unpaired (free) electrons, so it is very reactive. Radicals often form in a plasma.

When fluorocarbon compounds are broken down in a plasma, one important product is the CF_2 radical. This radical has two unsatisfied chemical bonds so it has a strong tendency to join many other such radicals and form chains called *polymers*. This kind of reaction is the basis of the plastics industry.

The polymer, polytetrafluoroethylene, chains of those CF_2 radicals mentioned previously, is one of the types of polymer that form in oxide etch plasmas. It is a form of Teflon® and it is very resistant to chemical attack. The polymers formed in etch plasma may not be quite as tidy in chemical composition as this, but they are still effective passivants.

The popular parallel-plate reactor (Figure 4-11) has been chosen to illustrate how an etcher accomplishes the objectives required by the process.

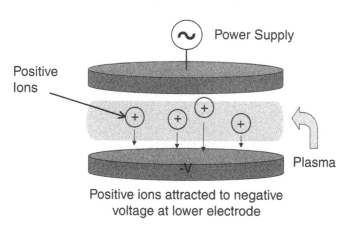

Figure 4-11: The Plasma in a Parallel Plate Reactor

The etch reactor operates at a low pressure. Standard atmospheric pressure is 760 Torr (14.7 psi). Etch reactors operate at pressures much lower than that, usually in the millitorr range (thousandths of a Torr). That may sound like a quite low pressure but at one Torr, there are still about 3.5×10^{16} atoms or molecules of gas in a cubic centimeter. That's over a million billion particles of gas. Clearly, there is still plenty of gas available to make the chemical reaction go.

Evangelista Torricelli

The favorite unit of pressure used by the semiconductor industry is the Torr. The name of that unit of pressure honors Evangelista Torricelli, who invented the barometer. Torricelli filled a long glass tube that was sealed at one end with mercury. Then, with his thumb over the open end of the tube, he upended it and immersed the open end in a bowl of mercury (Hg). The mercury dropped away from the sealed end a short way and fell no further. The column of mercury stood, in today's units, about 760 millimeters high. The column height varied slightly from day to day, but the average remained 760 mm. Today, the accepted value of standard atmospheric pressure at sea level is 760 mm of Hg or 760 Torr.

It is another one of history's surprising connections is that Signor Torricelli was Galileo's secretary.

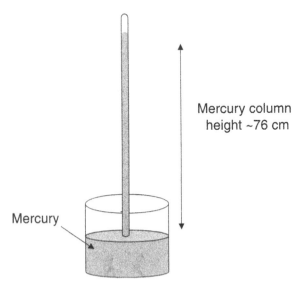

Figure 4-12: Torricelli's Barometer

A controlled amount of the desired gas is allowed to flow into the chamber. A special set of vacuum pumps continually removes the gas from the chamber. Recall that the products of reaction must be volatile for the etch to proceed so they are being continually pumped out of the reactor.

As shown in Figure 4-11, the wafer sits on one electrode. When the power is turned on, an electric field forms between two electrodes. The ions formed in the plasma are accelerated toward the wafer providing the bombardment energy to drive the etch reaction vertically downwards.

Key Point: The diffusion of gas molecules within the reaction chamber is the primary way that gas reactant reaches the wafer surface.

Figure 4-9 attempts to illustrate the "chemical soup" inside the etch reactor; everything is happening at once:

- Adsorbed reactants etch the oxide

- Bombarding argon (Ar) ions add energy to speed the rate of reaction

- Deposition of polymers for passivation of sidewalls and improved selectivity; vertical sidewalls are critical to this etch

- Bombardment physically knocks off (sputters) the deposited passivation from horizontal surfaces, exposing them to reactant

- Sputtering of exposed film is minor but exists in some etches

The remaining photoresist is removed by the plasma etcher prior to proceeding to the next step, etching the trenches. Photoresist etching or "ashing" was defined earlier in this section.

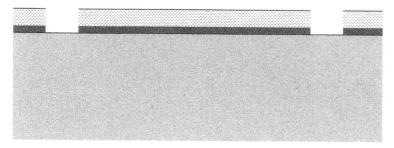

Figure 4-13: The Completed Hard Mask

Form Trenches in Silicon with Plasma Etch

Definition: *Trench Etch* is the process of etching trenches in the silicon substrate.

Purpose of Trench Etch: The trenches are shaped to exacting specifications so that, when filled with insulating oxide, they will provide a sufficient barrier to current flow between transistors.

Discussion

Silicon etch is done with chlorine and bromine. The trenches are etched primarily with chlorine. Bromine enters the process in later steps when additional rigorous requirements limit the use of the more aggressive chlorine.

The need for sidewall passivation is just as prominent for trench etch as for any application. The passivation determines the slope of the sidewall and it helps with other aspects of shaping the trench bottom and top edges.

The passivation produced at trench etch is very unusual in that it is not an organic material. The addition of oxygen to the silicon etch recipe produces a silicon-oxygen-chlorine byproduct that is not volatile. This compound is present in small amounts when similar chemical combinations are used for other applications, but here there is so much silicon to be removed that useful quantities of the compound are available. The amount produced is easily controlled by simply adjusting the amount of oxygen used in the etch recipe.

The inorganic passivant requires a wet dip in dilute hydrofluoric acid (HF) to remove it after processing.

Example of Silicon Etch Chemistry

$$Si(s) + Cl_2(g) + O_2(g) \rightarrow SiCl_4(g) + SiOCl_x(s)$$

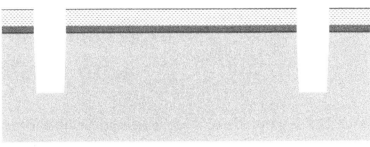

Figure 4-14: Completed Trench with Hard Mask Still in Place

In this example, silicon combines with chlorine and oxygen to form silicon tetra-chloride and silicon oxychloride. Fluorine is too aggressive to be used for silicon etch and produces an isotropic profile. Chlorine etches anisotropically but still has a high etch rate so it is preferred for this application.

Here is the completed trench. The shape will vary somewhat depending on the part being made, but this generalization shows the principal features.

The sidewall slope should be close to vertical so that space is conserved. Slight bottom rounding (not shown in Figure 4-14) is important to ensure that the trench is easily and completely filled with oxide, leaving no voids that might trap contaminants and cause a reliability issue.

The top edges of the trench are also slightly rounded (also not shown in figure) to minimize a current leakage problem. Sharp edges on a conductor cause the electric field to intensify and may create strips at the edges of the transistor which remain conductive even when the transistor should be turned off.

Fill Trenches with Silicon Dioxide

Definition: The *Trench Fill Process* consists of two parts. First, the trenches are lined with thermal oxide. Second, the bulk of the trench is filled with deposited CVD oxide.

Purpose: Oxide is the insulating material that provides the electrical isolation between transistors. Transistors are packed as closely together on a chip as they can be. Without good electrical isolation, they would electrically interfere with each other and the chip would not work.

Discussion

The two-step process for trench fill is necessary for several reasons. The thin thermal oxide liner, grown first, electrically stabilizes the silicon surface. It also rounds off the sharp corners, which create the electrical problems discussed earlier.

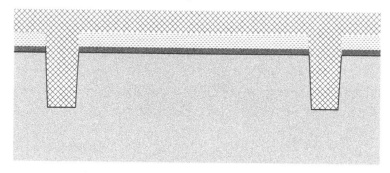

Figure 4-15: Oxide Deposited to Fill Trenches

121

The bulk of the oxide fill is done with a CVD oxide. Each type of oxide has different gap fill characteristics and produces oxides with varying properties. Void formation must be avoided to prevent trapped contaminants and production of parts with varying electrical properties. The nature of the fill oxide will also affect the stresses produced in the silicon at the trench sidewalls during oxide densification anneals and other thermal cycles in wafer fabrication. These stresses can result in increased pn junction leakage in the transistors. In addition, the characteristics of the fill oxide will affect the CMP step.

An example of a deposition technology that can fill trenches with good quality, void-free oxide is high-density plasma (HDP) CVD. HDP combines deposition with a small etching component in the process. The details of this technology can be found in some of the references in the Bibliography (Wolf, Ghandhi).

Refer to Chapter 3 for a discussion of thermal oxide growth and Section 3 of this chapter for CVD processes.

Chemical Mechanical Polishing (CMP) to Remove Excess Oxide

Definition: *Chemical mechanical polishing* is a process that gradually removes material from the wafer by a combination of physical abrasion and chemical action that is usually assisted by chemical reaction.

Purpose: The purpose of CMP is to remove material from the wafer surface without damaging the devices being built on the wafer. Planarizing (flattening) the wafer surface is required for some applications and a side benefit for others.

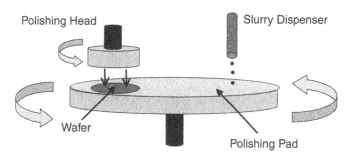

Figure 4-16: CMP Tool

Discussion

In the STI process, the surface of the wafer is quite uneven after the oxide trench fill. Recall that the nitride/oxide hard mask must be removed to uncover the areas where the transistors will be built. But first, the unwanted deposited oxide piled on top of these areas must be removed—and without removing the oxide in the trenches. CMP can do that effectively and leave the surface well planarized.

The need for a flat surface on the wafer has led to the use of CMP at several places in the chipmaking process. It has replaced plasma or wet etch, which was formerly the only film removal method available. Clearly, the chemical removal of a film will be nearly conformal and although many clever schemes were tried, etch could never do an adequate job of planarization.

The original use of CMP in wafer fabrication was to planarize the surface as an aid to photolithography. That application and several others will be discussed in later chapters.

Process Flow

1. Polish silicon dioxide/silicon nitride without breaking through the nitride
2. Post-CMP rinse/dry

Chemical mechanical polishing may seem like a very new and sophisticated technology but it is actually based upon a very old process. The fundamental process is used for lens grinding and has been in use for centuries. The use of CMP for polishing plate glass goes back a century or more. In the early years of the industry, the bare starting wafers needed for chip fabrication were polished using successively finer grades of abrasive particles.

In the 1960s, chemical-mechanical polishing was developed for silicon wafers. The introduction of CMP resulted in less damage to the crystal structure at the wafer surface than with abrasive-only polishing. Remember that devices are built in a very shallow layer at the wafer surface and this damage can easily harm device performance in various ways, notably by increasing junction leakage currents.

In the 1990s, CMP found its way into the wafer fabrication cycle, first to planarize irregular surfaces to assist photolithographic patterning of small features and then for several other applications as we will see later.

CMP Introduced to the Fab

Experienced engineers remember the first time that CMP was introduced to fab processing. Until that time, it was absolutely forbidden to touch the front of the wafer because it would kill the devices and reduce yield. It was a shock to hear that the wafer, covered with partially completed chips, was to be placed face-down on a grinding wheel and slurry would be used to remove the film. Everyone was sure that the entire wafer would be ruined by the polishing or by the huge number of particles left behind afterward. That the opposite result came about is very fortunate because CMP solves many processing problems in the fab.

CMP Principles: The CMP tool must remove the film as fast as possible without leaving a rough, damaged surface behind. As it turns out, the best way to do it often involves using a very small amount of chemical to react with a very thin layer of the surface of the film (dissolution). That helps the abrasive particles to cut and pull away microscopic particles of the film from the surface. In addition, some metals polish easier if their surface is oxidized; the oxide form of the metal is easier for the abrasive to remove without damaging the underlying wafer. In other cases, only DI water is used as the liquid component of the slurry. There are a wide variety of mechanisms involved and the technology is changing very rapidly. Nevertheless, the principal film removal mechanism remains the physical activity of the abrasive.

How It's Done: CMP is really tricky. There is a lot going on. First, here is a list of the main components of the CMP process:

1. The Pad
2. The Slurry
    a. The Abrasive Particles
    b. The Liquid Components

The liquid component of the slurry will typically contain deionized (DI) water, an acid, a base, an oxidizer or some combination of those four. The liquid often dissolves a so-called monolayer or two of material. It also carries the abrasive to the wafer and carries the particles away from the wafer. Let us not forget about heat—the polish rate is very temperature dependent. The liquid helps carry away the heat generated by the abrasive polishing.

The abrasive is also highly specialized. Many materials are available to use but these three dominate the practice today: silicon dioxide, aluminum oxide (alumina) and cerium oxide.

It may come as a surprise but the polishing pad is a very complicated and specialized material. Again, there are several choices for the pad material but polypropylene dominates the industry at this time. Enthusiastic experimentation is currently taking place and new pad materials could emerge on production tools at any time.

A typical CMP tool will have many process variables, often eighteen or more, that all have to be adjusted to optimize the process. Some of the most important are the amount of pressure pushing the wafer against the pad, the slurry flow rate, the pad speed, the percent solids in the slurry, and on and on.

Figure 4-17: Post-CMP Clean Tool

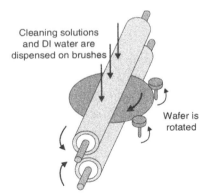

The post-CMP clean is almost as challenging as the CMP process itself. All of the particles that were deliberately generated during the polish must be removed. The liquids that are used for that purpose are much the same as the slurry but without the abrasive component.

The special characteristics of the particulate materials can be used to aid in removing them. For example, the STI polish is primarily an oxide polish. Recall that the oxide, or glass particles are good electrical insulators. That also means that they will hold a static electrical charge. When SiO_2 particles are placed in a basic solution, they take on a negative electrical charge. Ammonia, a familiar base, is a favorite component in the cleaning solution. If all of the particles have the same charge, they repel each other. In addition, the remaining oxide on the wafer surface takes on the same negative charge. All of this electrical repulsion makes it much easier to remove the particles; they do not tend to agglomerate and redeposit on the wafer, nor do they stick to the wafer surface.

The post-CMP clean is followed by a DI water rinse and hot nitrogen gas drying operation. The wafers are then ready for the next processing step.

Figure 4-18: Filled Trenches after CMP

Wet Etch Removal of
Silicon Nitride and Pad Oxide

Definition: *Wet Etch* is the use of liquid chemicals to remove material from the wafer surface.

Purpose: This step removes the remaining silicon nitride and the pad oxide from the wafer while leaving the underlying silicon undamaged.

Discussion

Wet chemicals such as phosphoric acid (for nitride etch) and dilute hydrofluoric acid (for oxide etch) will remove specific films with very high selectivity to other films. In other words, the nitride and pad oxide can be stripped off without damaging the underlying silicon substrate. That is very important because the surface of the Si wafer below the nitride is where the transistors are to be built. That surface must be kept in near-perfect condition.

The automated wet bench sequentially puts the wafers in acid baths and DI water rinse baths. Processing time is carefully controlled.

Automated tools are a major improvement to safety in the fab. Fab personnel are no longer required to constantly come in close proximity to the processing chemicals. The liquid chemicals can be very dangerous. For example, hydrofluoric acid (HF) attacks the bones without burning the skin so a person is often unaware of accidental exposure until some time has passed. Phosphoric acid is heated to about 180°C, and a favorite photoresist stripping solution, a mix of sulfuric acid and hydrogen peroxide to about 100°C. These acids and oxidizers are dangerous at room temperature but are even more so at high temperature.

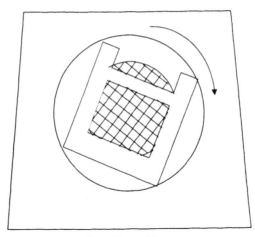

Spin-Rinse/Dryer
Wafers in carrier are rotated in an enclosed
chamber while water rinsed and air or nitrogen dried

Figure 4-19: Spin Rinser-Dryer

After processing, the wafers go into a spin rinser-dryer as shown in Figure 4-19. The tool spins the wafers, rinses them with DI water and then dries them in hot nitrogen gas.

Figure 4-20: Completed Oxide-Filled Trenches

The wafers may now proceed to the next major operation in the manufacturing process: the building of the transistors.

CHAPTER

5

Building the Transistors

Introduction

In this chapter, you will learn:

- How the transistors are built

- How the field effect transistor works

- The importance of high-k materials for the gate dielectric

- Pushing the limits of technology to make faster parts

- How the lightly doped drain (LDD) is formed and why it is needed

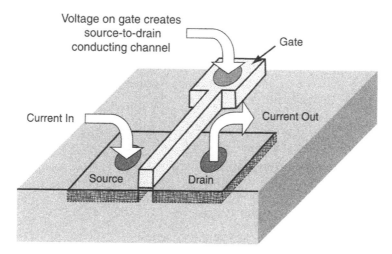

Figure 5-1: CMOS Transistor

Definition: A *Metal Oxide Semiconductor Field Effect Transistor (MOSFET)* is the fundamental electrical component of the CMOS chip. The action of the transistor is controlled by an electric field. As mentioned previously, the "metal" in the name refers to the earliest chips that used aluminum, instead of the polycrystalline silicon in use today, as the gate electrode. The oxide and semiconductor elements of the transistor are familiar to the reader by this time.

Purpose: The MOSFET is very versatile and offers many options to the circuit designer. For this discussion, the MOSFET is used as a switch.

MOSFETs and Other Transistors

In Chapter 3, it was pointed out that the "computer chip" would be the focus of this discussion because that type of chip clearly illustrates the function of a large percentage of the parts made today. It is important to bear in mind that there are many different kinds of transistors used in other types of chips and electrical components that will not be discussed in this text. The most notable example is the bipolar transistor that can be used for many applications. In fact, nearly all the chips made in the first ten years of IC manufacturing were based on bipolar transistors, the first kind of transistor invented. It can act as a switch, as an amplifier or perform other specialized functions.

Some titles for further reading are mentioned in the appendix.

Transistor Building Process Flow

1. Thin film formation
    a. Gate dielectric oxidation
    b. Deposit polycrystalline silicon
    c. Deposit nitride cap for hard mask

2. Poly gate formation using hard mask

    a. Photolithography process to pattern resist
    b. Trim resist in etcher
    c. Etch nitride hard mask
    d. Etch poly lines

3. Lightly-doped drain (LDD) formation
   a. Shallow, low-dose dopant implant
   b. Form spacers
      i. Deposit nitride
      ii. Etch spacers
   c. High-dose dopant implant
   d. RTP anneal

4. Salicide formation
   a. Sputter cobalt
   b. RTP reaction forming silicide
   c. Strip residual cobalt
   d. Anneal silicide (forms low resistance silicide)

Key Points in the Transistor Building Process

Each of the following points will be explained in the appropriate sections of this chapter.

1. High-k dielectric materials are needed to replace thermal oxide as the gate dielectric.

2. Resist trimming is often used to reduce the width of the poly line because some steppers currently in production cannot print a small enough feature.

3. A hard mask is sometimes required because resist trimming erodes the resist mask.

4. The lightly doped drain (LDD) process that produces a graded doping profile for the source and drain is required to minimize potential reliability problems.

5. Silicidation of contact surfaces is necessary for low resistance connections to the source, drain and gate. It also increases the electrical conductivity in those regions, speeding up the chip.

Discussion

Because the MOSFET is the heart of the chip, a short discussion of some of the particulars regarding it is in order. The focus of this discussion will continue to be on the n-channel MOSFET in the role as a switching component of a generalized computer chip. MOSFETs find uses in a very broad range of chip types but this discussion will be limited to its primary function in the digital computer field.

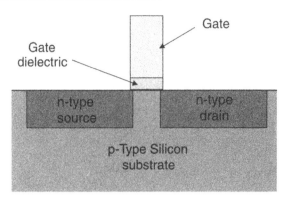

Figure 5-2: The n-channel MOSFET Cross Section

In this chapter, the MOSFET will be built. It is widely recognized as the workhorse of the electronics industry. Because MOSFETs will switch on and off at great speed, while using relatively little power, they can be used in digital logic circuits of all sorts. Today's computer technology has been built on the advances in MOSFET performance.

Although this discussion is limited to the use of MOSFETs in computer applications, the reader is reminded that these transistors are also widely used as amplifiers and other specialized components that are beyond the scope of this book.

Bear in mind that CMOS technology requires both n-channel and p-channel MOSFETs in combination. These pairs of transistors will be shown in the figures beginning with the last graphic in Chapter 5. For manufacturing purposes, the

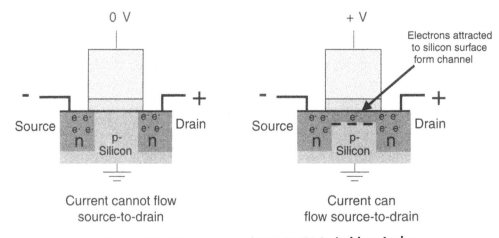

Figure 5-3: The n-channel MOSFET Switching Action

p-channel device is made generally the same way as the n-channel but the dopants are changed to reverse all of the polarities.

The transistor shown in Figure 5-3 is called an *n-channel transistor* since, in this form of the device, a positive voltage on the gate attracts electrons (negative charge carriers) to the silicon surface under the gate; the gate dielectric will not allow the electrons to pass through it so they accumulate at the interface. If the voltage applied is greater than what is called the *threshold voltage*, enough electrons will accumulate to form a conducting channel between the source and drain of the transistor. Now a current will flow between the source and drain, turning the switch "on." With no voltage on the gate, no channel is formed and the transistor is "off."

Digital Logic Note

In a simple digital chip like a DRAM, where the charge (or lack of charge) on a capacitor represents each bit of information, the OFF position is conventionally interpreted in digital logic as a zero (0) in binary arithmetic; the ON state is conventionally a binary one (1). Things get more involved with CMOS technology, where more than one transistor is involved in producing each digital bit of information. Logic circuits, for example, assign a low and high voltage level, relative to some reference voltage, to represent zero and one.

The name "field effect transistor" makes it clear that the electric field formed between the gate and substrate silicon is exactly what makes the device work. The electric field attracts the appropriately charged particles, either electrons or "holes," to form the conducting channel that, in essence, operates the switch. When the voltage to the gate is turned off, the electric field formed by the gate collapses, the charges flow away and the channel disappears.

Another name for the channel is "surface inversion layer" because, for our example, the presence of all the negative charges pulled into that very thin surface region inverts the ratio of holes to electrons there (majority/minority charge carriers). In other words, the electrons held in the channel fill the p-type holes with plenty left over to make the channel n-type. As long as there is a large enough voltage on the gate, the channel remains n-type and current can flow between the n-type source and drain because they are now connected by a continuous piece of n-type conducting material.

The width of the gate line or "stripe" is often considered the most important dimension on the chip. That dimension is the defining one for a specific "technology node." In other words, when someone says, "90 nanometer technology," they are referring to the width of the gate line. Some confusion does exist with the terminology because occasionally the actual channel length, a smaller dimension, is being referenced. This issue will be explained in more detail in Section 4.

Transistor performance is considered by many to be the driving force behind all the technological development associated with computer chips. Faster switching speeds means faster computers. What accounts for the speed of a transistor?

Several factors are important in the speed of operation of a transistor. The channel length is often the first thing that comes to mind. Since the channel electrically connects the source to the drain, the shorter the channel, the faster the chip. The principle at work here is that a short channel will form faster than a long one when the voltage is applied to the gate. In addition, a shorter channel will have lower resistance, an important circuit speed factor. Of course, the channel can become too short and then it will no longer be able to prevent current from flowing from source to drain when the gate voltage is turned off.

The gate stripe was about 10 microns wide in early chips. Now it is less than 0.1 microns (100 nm), down by a factor of 1000!

Another factor that is pushing the limits of technology is the thickness of the gate dielectric. Reducing the electrical resistance of all parts of the transistor and all of the interconnections on the chip, too, will increase chip speed. Reduction of parasitic capacitances in the circuit is a speed factor that will be explained in two upcoming sections.

A Perspective on Chip Speed

Not all chips need to be state-of-the-art with very small gate dimensions and very thin gate dielectrics. Lower performance technologies are perfectly adequate for many applications.

Maximum performance is required for high-end computers and servers. Lower performance is fine for telephones, laptop computers and portable electronic devices, the routine "slow" logic applications. Lower speed means lower power consumption, longer battery life, and comparatively lower price.

Thin Film Formation

2.1 Gate Dielectric Oxidation

Definition: The *Gate Dielectric* is a layer of electrically insulating material sandwiched between the gate poly and the silicon substrate. To date, it has been made almost exclusively of silicon dioxide.

Purpose: The gate dielectric insulates the gate electrode from the silicon substrate.

Discussion

The gate dielectric that is currently used in manufacturing is made of thermally grown silicon dioxide (silica). It is typically grown in an RTP system using oxygen and differs somewhat from the thermal oxide discussed in Chapter 3. Small amounts of nitrogen are often incorporated into the film to improve long-term reliability and reduce leakage currents.

The gate dielectric is the thinnest film on the wafer and has the most critical requirements for composition integrity and thickness. Thermal oxide is amorphous. The three-dimensional structure is a random combination of tetrahedra and triangles made of silicon dioxide with extra oxygen ions joining the silica polyhedra. This complicated structure is explained in Ghandhi (see the Bibliography).

The gate oxide thickness has been scaled down over the years to improve chip performance. With a thinner gate oxide, more conducting charges are attracted into the channel, reducing its resistance and speeding up the chip. Oxide thicknesses in the region of 20 Å, or 2.0 nm, are currently in use. These films are only a few atomic layers thick, approaching the point where even silicon dioxide is no longer a good enough insulator and too much current can leak through between the gate and the substrate.

Definition: k is the dielectric constant, a dimensionless quantity that serves as a way of comparing one of the insulating characteristics of various materials. k is defined as the amount that the capacitance increases, compared to vacuum, in a capacitor when that material is used as the dielectric. It must be determined experimentally for each material. By definition, vacuum serves as the reference. The value of k for vacuum is unity (1), the lowest possible value for k (k for air varies only slightly from vacuum).

Definition: *High-k Gate Dielectrics* increase the amount of charge that will accumulate in the channel for a given dielectric thickness and applied gate voltage, lowering the transistor's "on" resistance and speeding up the chip.

High-k materials improve on silicon dioxide by allowing the use of a thicker film without changing the performance of the transistor. Thicker films will reduce or eliminate leakage currents through the dielectric. Although some chips will tolerate a moderate leakage current, it increases the power consumption, so it is always undesirable.

To illustrate, silicon dioxide has a k of about 4. If a suitable replacement material could be found with a k of 8, the same electrical performance would be seen in the chip but the gate dielectric could be twice a thick. That would result in reduced power consumption because of reduced leakage current.

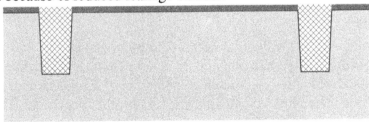

Figure 5-4: Gate Dielectric on Wafer

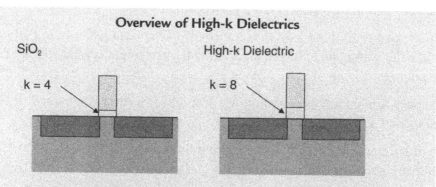

Overview of High-k Dielectrics

Doubling the k allows the thickness of the dielectric to be doubled.
The transistor performs the same but with less current leakage between
the gate and the silicon substrate

Figure 5-5: High-k Dielectric vs. Silicon Dioxide

There are many high-k materials with some of the required properties to be potential candidates to replace silicon dioxide as a gate dielectric. These materials are mostly oxides or silicates of certain metals.

Unfortunately, most of them have a serious failing. One common problem is that some will react with the silicon wafer or polysilicon gate to form a layer of silicon dioxide at the interfaces that excessively lowers the overall *k* of the film.

Another frequent problem concerns the electrical charge that is usually formed at the wafer/dielectric interface. A layer of charge here is equivalent to having a permanent, built-in voltage on the gate. High levels of charge which occur with some dielectrics affect the threshold voltage of a transistor very strongly but low levels of charge can be tolerated. This surface state charge, as it is called, results from the discontinuity in the chemical bonding at the interface between the silicon and various dielectrics.

At the silicon/silicon dioxide interface, a relatively small amount of positive charge is always formed. When MOS technology was first introduced, a great deal of work was done to develop a process that would consistently produce the same low level of charge so that the device engineers could correct for that characteristic. This well-understood and well-behaved silicon/oxide interface is not proving easy to duplicate with high-k films.

High-k dielectrics that have shown promise include the oxides and silicates of hafnium and zirconium but many other materials are being investigated and much work will be needed to accomplish this major change in the technology.

2.2 Polycrystalline Silicon (Poly) Deposition

Definition: Polycrystalline Silicon is a form of silicon that is deposited on the wafer. Unlike the wafer and epitaxial silicon, which are single crystal forms of silicon, the poly is made up of grains. Each grain is a crystallite, a tiny crystal of silicon.

The poly film thickness is commonly around 2500 Angstroms thick. It is deposited using LPCVD technology.

An important thing to remember about poly is that it will be doped to make it a relatively good conductor of electricity.

Polycrystalline Silicon Film Characteristics

- Standard LPCVD deposition at about 600°C
- Grains become smaller as deposition temperature is decreased
- At some lower temperature, amorphous silicon is formed (no crystal structure)

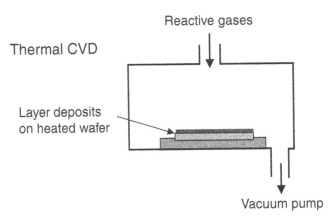

A process using chemical reactions in
gases to deposit layers of solid material

Example reaction: SiH_4 (gas) $\rightarrow$ Si (solid) + H_2 (gas)

Figure 5-6: Chemical Vapor Deposition

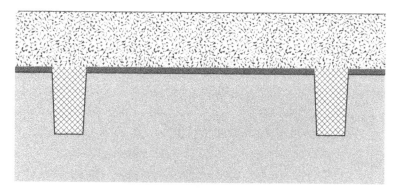

Figure 5-7: Poly Deposition

Notes on Gate Electrode Materials

Polycrystalline silicon is preferred for the gate material for several reasons. It will withstand high temperatures seen in succeeding processing steps. In addition, the poly/gate oxide interface is well understood, and conventional fab processes produce consistent results.

The poly is doped heavily for low resistance to current flow. Doped poly is fine by itself as the gate material for lower performance chips. High performance chips use heavily doped poly with silicide on top. This application is discussed in Section 5.

The first gates were made of a metal, usually aluminum, but molybdenum was used a little, too. Section 3 has more to say about this technology. Poly replaced metal for reasons mentioned above. However, metal gates may return to the industry soon for use in very high- speed chips. Development work is underway.

2.3 Nitride Cap Deposition

The nitride cap is standard PECVD nitride, the same hard mask process as found in Chapter 4.

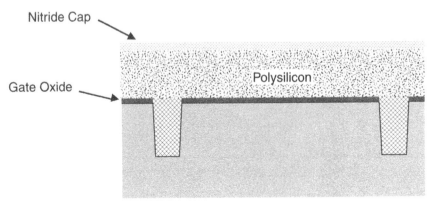

Figure 5-8: Completed Gate Film Stack

Poly Gate Formation

Poly Gate Formation Process Flow

1. Photoresist Patterning
2. Plasma Etch
    a. Resist Trim
    b. Nitride Hard Mask Etch
    c. Poly Etch

3.1 Photoresist Patterning

The first step in the poly gate formation process is resist patterning at photo. See Chapter 3 for a review of photolithography.

Pushing the limits of technology is vividly illustrated here. Producing today's fast chips with tiny geometries requires going beyond the current capabilities of photolithography in manufacturing. The details are discussed in the next section under resist trimming.

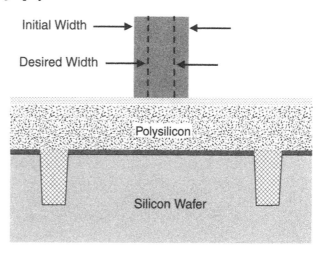

Figure 5-9: Gate Film Stack with Patterned Resist

3.2 Plasma Etch

Definition: *Resist Trimming* is the process of using an oxygen plasma to etch the patterned resist and thereby reduce the size of the pattern features.

Purpose: In order to make shorter channel lengths, and therefore faster parts, the poly linewidth dimension must be reduced.

Discussion

Resist trimming has been used for many years and is still going strong in production today. New stepper technology continues to innovate, always improving the ability to pattern smaller features. But onrushing demand for shrinking dimensions and faster chips requires a full engineering bag of tricks. Resist trim is an important one.

Here is a simple example of the need for resist trimming:

Many steppers use a light wavelength of 248 nanometers (248×10^{-9} m or 248 billionths of a meter long). The best lens systems can theoretically resolve a feature that is one-half of a wavelength in size or, in this case, 124 nm. For reference, that may also be expressed as 0.124 microns. That is the absolute limit of the feature size that can be produced in the resist. New scanner and stepper technology has found some other clever ways of pushing this boundary, but for the sake of this simplified example, those innovations will be overlooked.

How can a gate poly line of 90 nm be made then? It is done by fooling Mother Nature and using the etcher to etch away the resist lines until they match the desired dimension. Of course, the resist is badly eroded by the trimming so it cannot serve as a proper mask for etching the poly. The etch masking problem is solved in the same way that it was for STI discussed in Chapter 4: with the use of a nitride hard mask.

Update on Stepper Technology

New stepper technology using the shorter wavelength of 193 nm is increasingly available today for state-of-the-art devices at the 90 nm node. Steppers using 157nm technology are slowly entering production. There have been large difficulties in pushing the technology toward the X-ray wavelengths of light, mostly as regards the complexity of making production-worthy photoresist and/or lens systems with sufficient capabilities.

Great strides are being made in other ways. Steppers employing phase-shift masks and the new immersion technology, where the stepper objective lens is immersed in liquid that contacts the wafer surface, have greatly improved photo's capabilities. Next generation lithography (NGL) is a program to develop a lithography technology utilizing electron beams (e-beam).

Even with all of this incredible technological progress, resist trimming is going to be with us for a while.

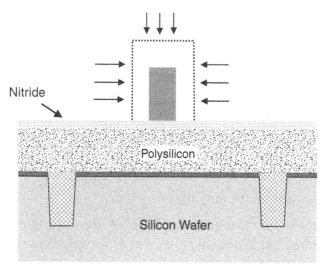

Figure 5-10: Resist Trimming Sequence

The resist trimming is done in a plasma etcher, most often using an oxygen plasma. The oxygen plasma oxidizes (burns) the resist, primarily forming the volatile products carbon dioxide and water vapor.

The wafer enters the plasma etcher and the resist is trimmed. Next, the nitride hard mask is etched, followed by the poly line. It is all done in the same chamber

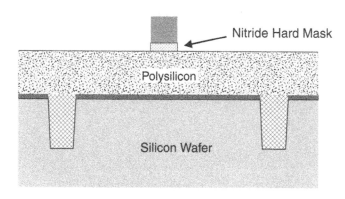

Figure 5-11: Nitride Etched to Form Hard Mask

most of the time. In other cases, a cluster tool may be employed. Cluster tools are machines equipped with multiple processing stations such as etch reactors and deposition chambers, that allow the wafers to go through multiple processing steps without breaking vacuum. Keeping the wafers under vacuum eliminates most sources of particulate and atmospheric contamination.

Silicon nitride is etched using the etch chemistry discussed in Chapter 4. The nitride is quite thin so the trimmed resist mask is sufficient to mask the etch.

The poly etch is performed using the chlorine/bromine etch chemistry discussed in Chapter 4. This is the most critical etch in the entire process. Two of the most important requirements of the etch are:

- Etch-stop on thin gate oxide
- Vertical sidewalls on the poly lines

The gate oxide is so thin that great care must be exercised to not break through it. The preservation of the oxide film on either side of the gate is for the protection of the silicon wafer surface directly underneath. Since the source and drain will be formed there, the silicon surface should not be damaged.

The poly features should have vertical sidewalls to be sure of producing the narrowest useful poly line. The poly is a conductor. It must have a sufficiently large cross-sectional area so that it minimally resists current flow through it. However, it must be narrow so that it forms the shortest possible channel. So a well-shaped, rectangular cross-section is desirable.

Any remaining resist is stripped off after poly etch.

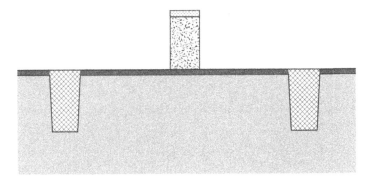

Figure 5-12: Completed Poly Gate Etch

Source/Drain Formation

Process Flow for Lightly Doped Drain (LDD) Formation

1. Shallow Implant

2. Spacer Formation

a. Deposit silicon nitride

b. Etch silicon nitride

3. High-Dose Implant

4. Anneal

4.1 Introduction

Definition: The *Source* and *Drain* are the names of the doped regions at either end of the channel. They are created by appropriately doping the silicon substrate in those regions to make it highly conductive, in accordance with the design specifications of the transistor. It is always necessary to add extra dopant to change the dopant profile of the desired region from p-type to n-type or vice-versa. Figures 5-1 and 5-2 in this chapter show the source and drain.

Purpose: The source and drain are two of the electrodes in the transistor. Electrical connections are made to these electrodes, as well as to the gate and substrate, making the transistor a part of the circuitry of the chip. As discussed earlier in the chapter, current flows between the source and drain when a suitable voltage is applied to the gate.

Definition: A *Lightly Doped Drain (LDD)* has specially-shaped doped regions forming the source and the drain that optimize the performance and reliability of a chip. In fact, both the source and drain have a lightly doped region, in spite of the name.

Purpose: The LDD helps to solve a reliability problem that was encountered as the channel length became shorter with the continued shrinking of the chip.

Discussion

The LDD is a two-implant process that positions lightly doped areas at either end of the channel. The more heavily doped areas are needed to minimize the resistance between the ends of the channel and the source and drain contacts and also to give low contact resistance to the metal connections. The light doping at either end of the channel reduces the acceleration of electrons or holes being drawn into the drain from the end of the channel. Before the adoption of this innovation, some electrons were accelerated to high enough velocities that they would damage the gate dielectric. This effect was known as the "hot electron" problem. The problem is much less serious in the case of holes in p-channel transistors.

Definition: A *Self-Aligned Gate* is the name of the source, drain and gate electrodes that were formed using the *Self-Alignment Process*. The self-alignment process creates the source and drain on either side of the gate in precisely the correct positions.

Key Point: The trick to making a self-aligned gate is to etch the poly line before implanting the source/drain. The poly line acts as a perfectly placed implant mask, so the dopant goes exactly where it should.

Purpose: The self-aligned process puts the three components in exactly the right place relative to each other. The speed and reliability of the chip are maximized in this way.

Discussion

Self-aligned gate is something of a misnomer. A more correct label might be "self-aligned source/drain," since the function of the process is to align the source and drain, not the gate. The name seems to be a throwback to the problem of aligning the gate early in the industry's history.

The positioning of the source, drain and gate electrodes is obviously very critical. In early aluminum gate technology, the gates were added after the sources and drains were formed. The alignment was done in the aluminum patterning step and some amount of misalignment was expected. As compensation, wide gate stripes were used to ensure that the whole source-to-drain region was covered. However, the extra overlap on one side or the other hurt the performance of the chip by slowing it down.

4.2 Shallow Implant

The wafers first go to ion implantation for the shallow implant shown in Figure 5-13. For technologies with gates in the region of 100 nm, the depth of the shallow implant is about 30 nm or less into the silicon substrate, which may also be expressed as 300 Angstroms to allow us to compare to an adjacent structure. This implant layer is about 10% of the thickness of the gate poly which is around 2500 Angstroms thick. But it is over ten times thicker than the gate dielectric at 20–25 Angstroms.

The position of the dopant species is very well defined at this point. The anneal step later in the process changes the shape and slightly expands the boundaries of the dopant.

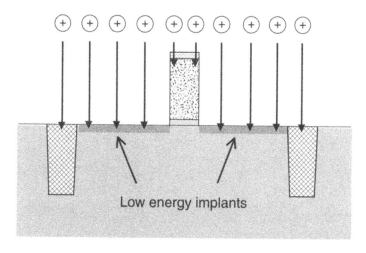

Figure 5-13: Shallow Implant, Self-Aligned Gate
(A thin screening oxide on the silicon omitted for simplicity)

4.3 Spacer Formation

Definition: A *Spacer* is an implant mask.

Purpose: The spacers cover narrow regions on either side of the poly line, blocking the high-dose implant that forms the bulk of the source and drain. The small shallow-implant regions that are protected by the spacers, called the source/drain extensions, contact the ends of the channel.

This clever device is a simple solution to the need for LDD's. Nitride or oxide will both work as the spacers but nitride is most common. In either case, the spacers are permanently left in place.

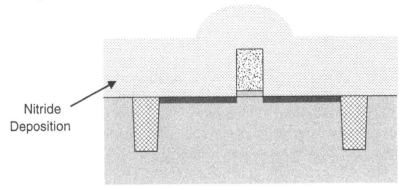

Figure 5-14: Conformal Nitride Film

A silicon nitride film is deposited next. The nitride layer follows the contours of the wafer surface and is about the same thickness as the poly. As shown in Figure 5-14, the conformal film will be twice as thick at the abrupt steps found at the edges of the poly lines. This thickness variation occurs at just the right spot and is the key ingredient in making the spacers.

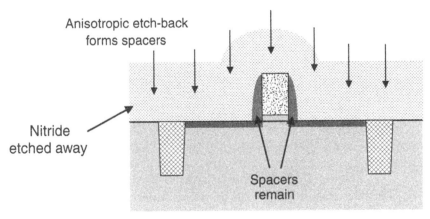

Figure 5-15: Spacer Etch

The spacer etch is a rather simple one. After the nitride is deposited, the wafer can immediately go to the etcher. There is no patterning involved. The one important factor in spacer etch is the uniformity of etch. The film must be removed from all parts of the wafer at exactly the same rate so that all transistors have the same size and shape of spacers.

4.4 High-Dose Implant

The high-dose implant follows the spacer formation. This implant is necessary to create a low resistance path between the ends of the channel and the source and drain contacts and to assist in forming low resistance contacts for connections that will be made to the source and drain.

In Appendix A, some of the doping characteristics of silicon are discussed. Recall that with dopant concentrations in the region of one part per million that the conductivity of silicon rose several orders of magnitude relative to intrinsic silicon. Typically, maximum conductivity is needed for the source and drain so high doping concentrations will be used.

Arsenic is the preferred dopant for n-type source/drain because it is a slower diffuser in silicon than phosphorous. Boron is the only viable dopant choice for making p-type material; unfortunately it diffuses just as fast as phosphorus so processing adjustments must be made.

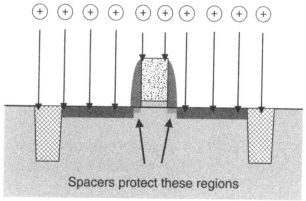

Figure 5-16: High-Dose Implant

4.5 Anneal

Definition: *Anneal* is a high temperature treatment that is used, in this case, to repair silicon crystal damage, activate dopants and distribute dopants within the silicon.

Annealing is commonly done in a rapid thermal processor (RTP). The process is called a *spike anneal* because the wafer is not held at the high temperature but rather heated to a peak and then quickly cooled. The temperature ramping rate is one of the critical process parameters. The peak temperature may exceed 800°C.

Before annealing, the dopants are electrically inactive. During implant, the silicon crystal lattice is broken up and the dopants and silicon atoms are randomly ordered. The anneal reconstructs the crystal with dopant atoms substituting for some of the silicon atoms in the lattice. Once this orderly structure is produced, the electrons or holes are released from the dopant atoms and are able to move independently within the solid. Now the dopants are said to be activated.

As shown in Figure 5-17, the dopants move somewhat from their original positions creating a changed dopant profile. This natural diffusion is a well-understood phenomenon and must be carefully controlled. It is a good way to shorten the channel to a small degree.

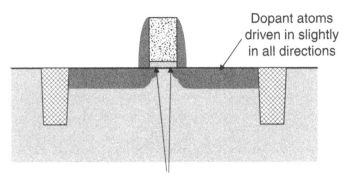

Dopant atoms driven in slightly in all directions

Lightly doped S/D *extensions* just under edges of gate

Figure 5-17 RTP Source/Drain Anneal

Dual-Doped Poly

Early in the history of the industry, the gate poly was doped heavily n-type for both n-channel and p-channel transistors. The thinking at the time was that the higher conductivity of the n-type versus p-type material would improve the speed of the chip. Now, in high-performance chips, n-channel gates are n-doped and p-channel gates are p-doped. In CMOS technology the n- and p-type transistors work in pairs called *complementary pairs*. Because of the effect on the transistor's threshold voltage, a PMOS transistor works better if the gate is p-doped. The Bibliography contains some suggestions for further reading on the functioning of semiconductor devices.

Salicide Formation

Definition: A *Silicide* is a chemical compound made of silicon and a metal.

Purpose: Silicides have been used for many years because they improve the conductivity of the gate electrode and reduce the contact resistance, speeding up the chip.

Definition: Polycide is the name of a stack of two films, silicide on polysilicon.

Discussion

The use of the polycide film stack gives the best of both worlds. The poly/gate oxide interface is well-understood and robust. The poly is easy to process and tolerates high temperatures. But the poly, even heavily doped, is not an exceptionally good conductor. Many silicides have resistivities that are much lower than doped poly; their use will significantly increase chip speed. And the silicide will tolerate high temperatures encountered in processing, too.

Definition: *Salicide* is the short name for self-aligned silicide. In this process, the silicide material only forms in exactly the right places in the transistors.

Definition: *Sputtering*, a common type of *Physical Vapor Deposition (PVD)* is the process of physically bombarding a target source with argon ions, which knock off atoms from the target, some of which find their way to the wafer and stick to it.

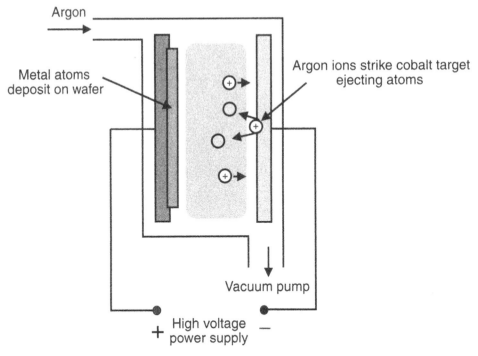

Argon

Metal atoms
deposit on wafer

Argon ions strike cobalt target
ejecting atoms

Vacuum pump

+ High voltage −
power supply

Figure 5-18: Sputtering

Historical Notes on Polycide

For many years, the resistances of poly gates and interconnections was reduced by forming a layer of a silicide on the poly before the gate mask was patterned. In these so-called polycide layers, the very low resistance silicide on the poly brings the resistances down to much lower levels than is possible with poly alone.

The tungsten or titanium silicides which were commonly used can be formed by first depositing a thin film of metal on the poly. Then in a high temperature step, reaction between the metal and some of the poly produces the silicide layer. More commonly, however, silicide films have been deposited over the poly using CVD processes.

In current high-performance chips, silicide layers are used on the sources and drains of the transistors as well as on the gates. Improved electrical conduction in the film is a major benefit. Another significant improvement is the reduction in contact resistance. Typically, cobalt or nickel silicides are now preferred and are not formed until after the sources and drains have been implanted.

5.1 Sputter Cobalt

Cobalt is a widely used metal in current silicidation processes.

Notice that the sputtered cobalt covers the whole wafer but the only places that silicide will form are the exposed silicon source/drain/gate areas. The silicide will not form on the oxide or nitride. The silicide is truly "self-aligned."

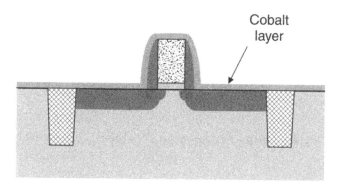

Figure 5-19: Cobalt Sputtered on Transistor

5.2 RTP Reaction Forming Silicide

The silicide will form when the cobalt reacts with the exposed silicon. All that need be done is to raise the temperature of the wafer to the reaction point and leave it there long enough for the reaction to happen. However, high temperature is a risk to the structures already completed on the wafer.

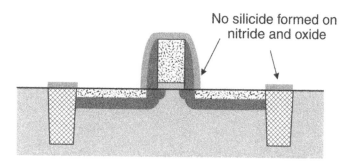

Figure 5-20: Silicide Formed on Source, Drain and Gate

The rapid thermal processor comes to the rescue. The value of the RTP is especially important at this time because the source/drain implants are already in place and have been annealed. Those structures would be damaged if subjected to a high temperature for too long a period.

5.3 Strip Residual Cobalt

The unreacted cobalt must be removed since it is a conductor and would short out all of the transistors if it were left on the chip. A wet etching procedure is used. The combination may contain an acid or ammonium hydroxide, hydrogen peroxide and DI water or some similar concoction.

5.4 Anneal the Silicide

Definition: *Annealing* is a high-temperature treatment that rearranges the structure of a material.

Purpose: The silicide annealing treatment lowers the resistivity of the cobalt silicide film.

One more step remains in the MOSFET making process. The silicide requires a high-temperature treatment called an *anneal*. This treatment is usually done in an RTP system. Anneals for silicides often take only about one minute at around 800°C. The anneal rearranges the atomic structure of the material, creating a lower resistance film, often adjusting the metal to silicon ratio.

The wafers may now proceed to the next major operation in the manufacturing process, making electrical connections to the transistors.

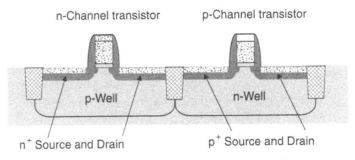

Figure 5-21: The Completed CMOS Transistors

CHAPTER

6

First Level Metallization

Introduction

In this chapter, you will learn:

- Why barrier layers are needed
- The importance of planarity
- Connecting the transistor to the rest of the circuit
- Metallization technology
- The importance of low-k dielectrics

Objective of First Level Metallization*:* to electrically connect the source, drain and gate electrodes of each transistor to other elements of the chip circuit.

First Level Metallization Process Flow

1. Nitride and oxide depositions
   a. PECVD SiN barrier layer
   b. CVD BPSG or PSG
2. CMP planarization of oxide
3. Pattern and etch contact holes
4. Tungsten plug process
   a. Deposit Ti/TiN barrier/glue layers
   b. CVD tungsten
   c. CMP excess tungsten

5. Intermetal dielectric (IMD) and Trench Etch

   a. Deposit IMD low-k dielectric

   b. Pattern and etch trenches

6. Copper process

   a. Ta/TaN barrier layer

   b. Sputter

   c. Electroplate

   d. CMP excess copper

   e. SiC barrier layer deposition

Key Points in the First Level Metallization Process

1. Barrier layers are needed to prevent a variety of problems

2. Adhesion or "glue" layers are used for making the films stick together

3. Planarization is required by photolithography technology

4. Tungsten CVD is required to fill high aspect ratio contact holes

5. Low-k dielectric materials are replacing deposited oxide as the intermetal dielectric (IMD)

6. Copper wiring for interconnections is required for state-of-the-art devices

Each of these points will be explained in the appropriate sections of this chapter.

Discussion

First level metallization has changed drastically in recent years. This processing step is dominated by new technology and new materials.

For many years, electrical connections to the source, drain and gate of each transistor were made with aluminum, the same aluminum film that was patterned and etched to form the interconnecting wiring between the devices on the chips. This aluminum technology is still widely used today for earlier generation chips. Indeed, many chip designs enjoy decades of continued use (see Figure 6-14).

As device dimensions decreased over the years, the depth of the pn junction also decreased. With shallower junctions, a problem called *junction-spiking* arose. When aluminum comes in contact with silicon, some of the silicon will dissolve in the aluminum; this phenomenon forms what is called a *solid solution*. The aluminum/silicon alloy will, in turn, move into the areas voided of silicon. With shallow pn junctions, these "spikes" could be long enough to penetrate through the junction, shorting it out. The shorts will prevent the chip from operating.

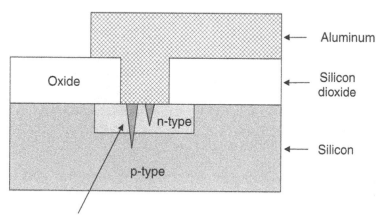

Aluminum alloy spike shorts through pn junction

Figure 6-1: Junction Spiking

The problem was first addressed by using aluminum with a small amount of silicon added to it (0.5–2%, typically). The idea was that if the aluminum was already saturated with silicon, it would not dissolve silicon from the contact areas. This worked to some degree but as junction depths became still shallower, it became necessary to take further steps. Soon a thin diffusion barrier layer was deposited on the wafers prior to aluminum deposition to completely prevent contact between the aluminum and the silicon.

Another need for barrier layers is the prevention of the movement of mobile contaminants through the chip. These contaminants can migrate through silicon and the films formed on the chip and could ruin the chip. More on this topic will appear in the following sections.

The thickness of the aluminum metal film was about one micron in 1970. Only one layer of metal lines was needed to connect all of the circuit elements at that time. The complexity, device density and small feature sizes characteristic of today's chips requires many layers of metal to make all of the connections. For these multilevel metal chips, the thicknesses are nearer to 0.5 micron for the lower levels and around one micron for the top levels where relatively high currents are carried around the chip. However, the metal line-widths have shrunk to one one-hundredth of their 1970 width. That shrinkage exacerbated the following problem.

Aluminum and other metals suffer from a problem called *electromigration*. The electrical current flowing through the aluminum wire will sometimes carry a few atoms of the aluminum right along with it. Enough atoms can be moved to produce voids and, eventually, complete breaks in the wires. It is interesting that a flow of electrons can move large numbers of metal atoms around. In the case of aluminum, one atom weighs about 50,000 times as much as one electron! The damage done by a hurricane or tornado is a similar phenomenon; wind is predominantly nothing more than fast-moving nitrogen and oxygen molecules, individually tiny compared to the huge structures that can be demolished by their cumulative energy.

Nitride and Oxide Depositions

2.1 Nitride Deposition

Definition: A *Barrier Layer* is a film or stack of films that acts as a barrier against the diffusion of atoms or molecules through it.

Purpose: Barrier layers perform a variety of functions. For example, they can be used to block silicon atoms from dissolving in aluminum; they will keep mobile ion contamination from migrating through the chip, protect the completed wafer from air and moisture in the atmosphere and may also act as a glue to hold two stacked films together.

Discussion

Silicon nitride is a widely used barrier layer and is also incorporated into combinations of films used as barriers or glue layers. The use of nitride as a hard mask and an implant mask have been discussed previously. It is a dielectric material that is critical to chipmaking.

The completed chips are often protected by a barrier of PECVD silicon nitride. The PECVD technique produces a nitride layer containing a lot of hydrogen, unlike the LPCVD Si_3N_4 discussed earlier in Chapter 4. The chemical formula for this film is often written as SiN. The PECVD nitride is still a tough dielectric and perfectly capable of doing the job of a barrier layer, but may be deposited at a lower temperature. If not for the lower temperature deposition, it could not be used because the higher temperature method would cause destructive chemical and metallurgical reactions in the chip wiring. After the first level of metal wiring is in place, no temperatures higher than about 400°C can be used.

One of the applications for barrier layers comes from the use of copper technology for the metal interconnects. The danger of copper contaminating the devices will be discussed in the upcoming section on copper deposition.

Another important task for the barrier layer is to keep the dopants in the doped oxide from migrating into the underlying layers and changing the doping profiles (review Chapter 5) of the conducting regions of the transistors. The next layer to be discussed in the process is often a deposited oxide doped with boron and phosphorous. Some movement of dopants from this layer into adjacent regions is possible.

For a discussion of nitride deposition, review Chapter 4.

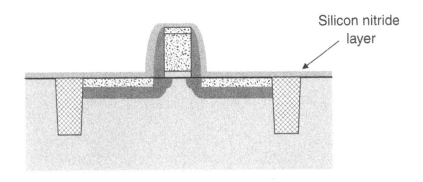

Silicon nitride layer

Figure 6-2: Nitride Barrier Layer

2.2 Oxide Deposition

Definition: The *Pre-Metal Dielectric (PMD)* is the thick insulating layer deposited to cover the transistors before the first level of metal wiring is formed.

Purpose: The pre-metal dielectric prevents electrical shorts between the conducting elements of the transistors and the wires that connect the transistors to the rest of the circuit. It is relatively thick to reduce electrical interaction between the wiring and the devices below. The primary concern is parasitic capacitance: the tendency for nearby conductors to interact when they are carrying a current. Parasitic capacitance delays the transmission of signals through the circuit wiring; it is discussed later in the chapter.

Discussion

The pre-metal dielectric is a CVD oxide that is doped with phosphorous or with both boron and phosphorous. The names of these films are phosphosilicate glass (PSG) and borophosphosilicate glass (BPSG). The heavily doped BPSG has been used for decades as the pre-metal dielectric because it will melt or "flow" at a comparatively low temperature. This property was used to get a smoother, flatter coating over the transistors, which made it easier to deposit uniform, conformal metal layers on the glass.

The use of CMP and tungsten CVD technology have eliminated the above concerns for state-of-the-art chips so PSG is a good replacement film. It has no boron so the risk of migrating dopants is reduced. However, BPSG is still widely used because it is a well-understood film.

There is a good reason to keep the phosphorous in the oxide film. A phosphorous-doped glass is an excellent "getter" of sodium. In other words, the film will trap and hold sodium ions that migrate into it. Sodium is a very common mobile ion contaminant, especially considering that it is in everyone's body and all over the skin. Our presence is dangerous to the wafer!

Review Chapter 4 and 5 for more information on CVD processes.

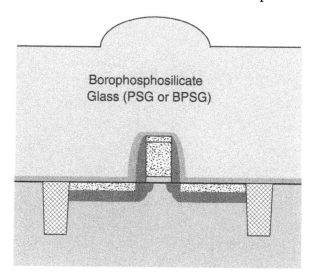

Figure 6-3: Cross-Section after Oxide Deposition

CMP Planarization

Definition*: Planarization* is the process of making the surface of the wafer as flat, or planar, as possible.

Purpose*:* The bumps on the surface of the wafer can be roughly as high as the depth of focus of the steppers in state-of-the-art tools. The stepper is unable to look down at the wafer surface and focus on all points of the wafer to print the next pattern—in this case, the contact holes. Without planarization, the holes at some points on the wafer would be out of focus, resulting in a poorly defined pattern—or no pattern at all—in the resist mask.

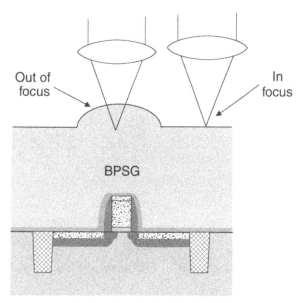

Figure 6-4: Illustration of Stepper Focus Problem

Discussion

Historically, planarization is the first application of CMP ever used on the front of the wafer after device structures were formed. Many industry professionals substituted "planarization" for "polishing" when using the term CMP because it was the first, and for a long time, the only CMP application performed in the fab. With so many polishes being used today, that terminology error has disappeared.

For a CMP review, see Chapter 4, Section 8.

The process is an oxide polish much like the one used for polishing the deposited oxide at STI. Planarity was one objective of that application, as it is here. In this case, the polishing operation only removes the bumps from the oxide layer and does not remove all of the oxide. Careful process control is important since it is necessary to leave sufficient oxide thickness for insulating purposes. Recall that the STI process polished off the excess oxide and stopped on the nitride layer underneath.

Review Chapter 4 for more details about CMP and oxide polishing.

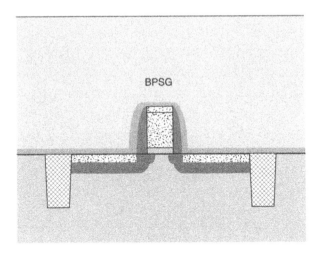

Figure 6-5: Planarized BPSG

Photo/Etch for Contact Holes

4.1 Contact Hole Photolithography

Definition: A *Contact Hole* is the hole that is etched completely through the pre-metal dielectric and the nitride barrier below it. The contact holes will be filled with metal, making electrical connection to the source, drain and gate of each transistor and to any other points on the silicon surface or conducting features where connections are needed.

See Chapter 3 for a review of photolithography.

Discussion

This patterning process is much like all of the other resist mask applications discussed previously. The photomask for contact hole etch is interesting because it is usually comprised of large arrays of tiny holes. It is not uncommon for the patterned resist to have openings to the underlying oxide that amounts to only about 1% of the total surface area of the wafer. There are millions of holes, it is true, but each one is very, very small.

Processing problems arise because of the extremely small diameter of the contact openings. The developer solution must be able to penetrate all the way to the bottom while dissolving the exposed resist, but surface tension inhibits that action. Surfactants (wetting agents) reduce surface tension in liquids and must be added to the developer to be sure that the solution dissolves all of the exposed resist to create the contact holes.

Surfactants must also be used for the DI water rinse. All of the residual developer solution must be rinsed out of the holes or a scum is left behind. The scum will inhibit the ability of the etch chemicals to contact the oxide and the pattern may not be properly etched into the film.

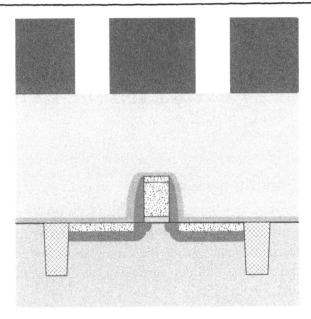

Figure 6-6: Contact Hole Resist Pattern

4.2 Contact Etch

Definition: A *High Aspect Ratio Contact (HARC)* is a contact hole that is very deep compared with its diameter. An 8:1 aspect ratio is often found in state-of-the-art contacts. *Aspect Ratio* is simply the ratio of the depth to the diameter of the hole. A high aspect ratio means that the depth of the hole is large as compared to the diameter of the hole.

Discussion

Contact etch is primarily an oxide etch which closely resembles the previous examples. It is generally fluorine-based with additional chemical components that produce the important sidewall passivation deposits. The process is often a two-step etch because the nitride film underneath the pre-metal dielectric must be removed from the bottom of the contact hole for the electrical connection to be made. Silicon nitride is etched with a similar chemistry to that used for the oxide glass.

The etch requirements are necessarily strict for these holes. The sidewalls must be vertical, or nearly so, because they are all so close together that any large degree of sloping could cause two holes to meet and short out when filled with metal. In addition, a sloped sidewall would produce a smaller opening at the bottom, reducing the size of the contact area, causing increased electrical resistance. So the amount of sidewall passivation produced in the etch chemistry is carefully regulated.

As one would expect, the high aspect ratio holes of such tiny diameters create chemical transportation problems: getting the reactant chemicals into the holes while removing the products of reaction from the holes, all the while passivating the sidewalls with large organic molecules. It is no easy task.

Reentrant profiles (undercutting) would also be a killer defect because the hole could not reliably be filled with metal. Any voids in the devices can trap air and moisture and become a reliability problem.

After etching through the nitride, it is important to stop the etch without removing much of the silicon. The surface of the silicon source, drain and gate electrodes is the contact point for the metal interconnects. In order for good electrical contact to be made, it is best if the silicon is undamaged. Only a very small amount of silicon erosion can be tolerated.

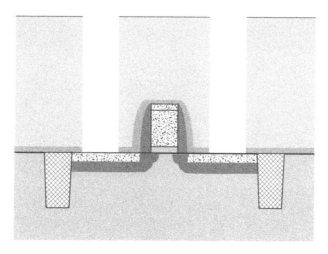

Figure 6-7: Completed Contact Etch

Historical Perspective on Aspect Ratio

It is interesting to note that while device geometries have shrunk dramatically with time, the thickness of the layers in device fabrication have decreased quite little.

The film thicknesses have remained about the same for circuit performance reasons. To maximize circuit speed and minimize power consumption, interconnection resistances and stray capacitances must be kept as low as possible. That requires rather thick insulating films separating conducting layers. Also, the metal for wiring has remained thick to minimize the circuit interconnection resistances.

The aspect ratio has increased dramatically because the features on the chip have been pushed closer and closer together while the film thickness has not changed much. Years ago, the diameter of the contact hole was about the same as the oxide film thickness giving an aspect ration of one or less. Today the film thickness is about the same but the hole's diameter has been reduced to less than a tenth of its earlier diameter. Eight to one aspect ratios are common now.

Tungsten Plug Process

5.1 Deposit Ti/TiN Barrier/Glue Layers

Definition: A *Glue Layer* or *Adhesion Layer* is a film that bonds well to two other films on the chip, holding them together.

Discussion

The junction-spiking story in Section 1 refers to this step in the process. Tungsten plugs have replaced the aluminum in the contact holes but barrier layers are as important as ever.

An early barrier material was a sputtered film of titanium/tungsten. This film was usually a mixture of 10% titanium and 90% tungsten sputtered from a target of that composition. Later, titanium nitride (TiN) was found to be a more effective diffusion barrier and has been very widely used ever since. Usually, a thin film of titanium is deposited before the TiN to reduce contact resistance. Titanium is a very reactive metal and will break through any thin native oxide film that grows readily on the exposed silicon contact areas.

Titanium also sticks to the oxide film very well. It is easy to forget that the films may be incompatible and delaminate (peel off). The need for a "glue layer" or "adhesion layer" is always a concern in the layering process.

Titanium is deposited using sputtering (PVD). There are both PVD and CVD processes for the deposition of TiN.

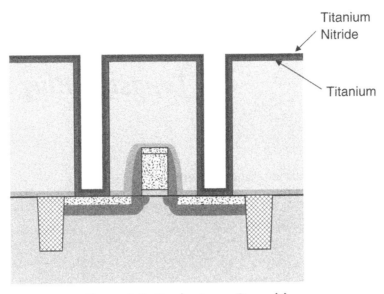

**Figure 6-8: Barrier Layer Deposition
(Not shown in later figures for simplicity)**

5.2 Tungsten CVD

It is very difficult to fill HARC holes with sputtered aluminum. Much work has been done on aluminum sputter technology in an attempt to overcome this limitation, but success has been very limited. The difficulty lies in the nature of sputtering. The naturally random scattering of Al atoms raining down in a PVD system will cover the surface of the wafer, but when small holes are to be filled, the method is unreliable as shown in Figure 6-9.

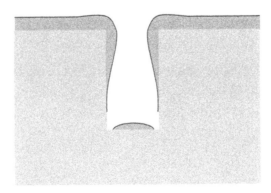

Figure 6-9: Metal Sputter Deposition Problems in Small Geometry Holes

The solution to the problem was to change the deposition method to CVD and change the metal to tungsten since there were many problems with aluminum CVD.

Tungsten (W) deposition is an LPCVD process. The film forms on all wafer surfaces and seems to grow up from the bottom and outward from the walls of the contact holes rather uniformly, regardless of the orientation or shape of the surface. The molecules of process gases move around rapidly in all directions and can easily enter small holes and react on all of the surfaces. The interesting characteristic of CVD-W is the formation of a seam that runs up the center of the plug and is clearly visible after CMP.

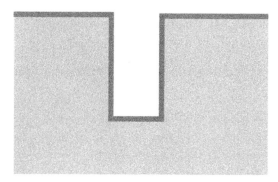

Figure 6-10: Tungsten CVD Film Deposition

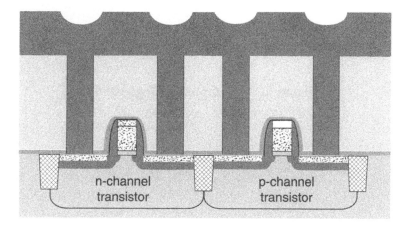

Figure 6-11: A Complementary Pair of Transistors with Tungsten Deposited to Fill Contact Holes

5.3 Tungsten CMP

The tungsten that fills the contact hole is called a *plug*. CMP removes the excess W from the deposition leaving the plugs behind. The polish must also remove the Ti/TiN barrier layer because it is also conductive and would cause all of the contacts to short out. Therefore, the polish stops on oxide, the pre-metal dielectric.

The tungsten polish is the first metal polish in the process flow. Metals present some interesting challenges and can be tricky to remove with CMP. A typical slurry contains an abrasive of silica or alumina in DI water. The inclusion of an acid and an oxidizer in the slurry is sometimes applicable but control of the surface removal can be more difficult with a metal film. There exists great variety in CMP slurries. Recall from earlier discussions of CMP that the primary film removal mechanism is abrasion no matter what liquid components are used in the slurry.

Metal defects such as spalling and voids from ripped off pieces of metal are a common concern.

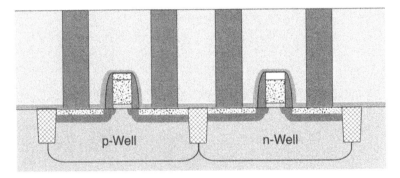

Figure 6-12: Completed Tungsten Plugs

Low-k Dielectric Process

6.1 Deposit Low-k Dielectric Film

Definition: *Low-k Dielectric Materials* are electrical insulators with relatively low dielectric constants. The lowest value of k is vacuum where $k = 1$; air is only slightly higher than 1. Silicon dioxide, the original intermetal dielectric (IMD), has a k of about 4. The objective is to find production-worthy materials with values below 4 and as close to 1 as possible.

Definition: *Intermetal Dielectric (IMD)* is the term for the insulating layers separating the interconnecting metal wiring patterns. It is also known as the *Interlevel Dielectric (ILD)*.

Even though this first low-k film could be considered to be part of a pre-metal dielectric film stack, it is often the same material as the later intermetal dielectric layers on the wafer, so the term IMD will be used here, as well.

Discussion

The IMD is an important insulator. It separates and insulates all of the interconnecting metal wires on the chip. There is more going on here than meets the eye, however. Even though the metal lines are separated by insulator, they still communicate with each other because they sense each other's electric fields. A parasitic capacitance can form at every place that the wires on different levels in the stack cross over each other or anywhere that wires come close together. These countless microscopic capacitors cumulatively slow down the chip because they must be charged and discharged every time a switching operation occurs.

Parasitic Capacitance

A capacitor, in its most fundamental form, is a charge storage device. Imagine two wires connected to the poles of a battery that are each attached to a coin. If the two coins are held with insulating material, facing and close to each other, without touching, charges from the battery will flow into them and be held there indefinitely; this arrangement constitutes a capacitor. Move the coins far away from each other and the charges will flow back into the battery. But disconnect one or both of the wires when the coins are close together and the capacitor will remain charged. Better not to touch them.

Using nontechnical terminology, parasitic capacitance is a phenomenon that occurs everywhere in the chip that conductors are positioned relatively close to one another. Although there is insulating material separating these conductors, they weakly interact. The voltage that can be felt between them attracts and holds a small number of charges. The effect is the same as having a small capacitor in the circuit. Many types of chips will have millions of crossing points in the wiring.

The problem with parasitic capacitance is that every time a transistor turns on, the electrical current in the wire carrying that signal must charge up all of the parasitic capacitors in the wire before the signal can pass. Many types of chips switch on and off millions of times a second. Needless to say, this issue slows the chip down.

The layers of wiring in a chip generally form a crisscross pattern, which helps to minimize the effect of each interaction point. In addition, the parasitic capacitance problem could be reduced by increasing the spacing between lines in the same level and by using thicker IMD layers. However, moving the lines farther apart would increase the chip size too much. In addition, there is a limit to the film thickness and the ability of etch to cut holes in the film for the interconnections. Further improvements in chip speed require lowering the k-value of the IMD.

The discussion of high-k dielectrics showed how those materials help more charges to form in the channel because of a strong interaction between the gate and the substrate surface. It is exactly the opposite effect that is desired here. Low-k materials improve the speed of the chip by minimizing the interaction between adjacent conductors.

Silicon dioxide was the standard IMD for many years but lower-k materials are replacing it. There are many candidates for replacing silicon dioxide as the IMD.

A wide variety of dielectrics with k-values in the 2.7 to 3.0 level have been investigated for many years. Some are inorganic, some organic and some are in between! Materials called *organo-silicate-glasses* (OSG) or *carbon-doped oxides* (CDO) with k-values of 2.7 to 3.0 have received a lot of attention. These materials are chemically similar to silicon dioxide, which makes them relatively easy to integrate into the manufacturing process. Some are deposited in conventional CVD systems that also offer the possibility of depositing silicon carbide barrier layers in the same process step.

Other types of CDO called *spin-on dielectric* (SOD) or *spin-on glass* (SOG) are applied by spinning them on like photoresist followed by a low temperature curing cycle. There are inorganic (carbon-free) versions of the spin-on materials, as well.

The all-organic dielectrics are basically special plastics. They are spun on and cured. They require much different processing than the silicon oxide-based materials and this has discouraged some companies from working with them, although one major company has been using one of them for several years.

Foams that are made from the kinds of materials described above offer great promise because their k-values approach 2. This low-k value is found by averaging the values of the bulk material and that of the gas in the pores or bubbles that form in the bulk material. Foams, however, present many process difficulties and do not stand up well to CMP. It appears that it will be some time before practical foams are developed.

Fluorinated silicate glass (FSG) has been widely used to replace silicon dioxide and has served the industry well down to the 100 nm–130 nm technology range. It is clear that new materials will be needed for the next technology node of 90 nm and beyond.

Currently, some of the organo-silicate-glasses and organics have entered production. The name "organo-silicate-glass" indicates that there are strongly organic components in the film. However, these materials can be etched with plasma

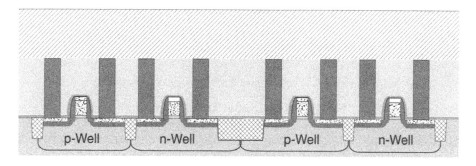

Figure 6-13: Low-k IMD Film Deposition

chemistries similar to those used for silicon dioxide. That characteristic, as well as their ability to stand up to CMP, make them very production-worthy materials.

Most every low-k IMD has a different deposition and etch process. This fascinating area of technological development is changing rapidly. The reader can enjoy many other sources of information to monitor these advances in chipmaking.

6.2 Trench Photolithography and Etch

Trenches formed in the IMD film are to be filled with copper in a "single damascene process" (see Chapter 7 for a more complete discussion of this terminology). This process forms conducting wires in much the same way that a decorative metal inlay is done. The trenches are formed by an etch process over the tungsten plugs that are to be electrically connected. The trenches are then filled with metal. The excess metal is polished off with CMP, leaving a conducting "wire" in the trench. Earlier technology formed the conducting wires by etching aluminum lines on top of the IMD (see Figure 6-14).

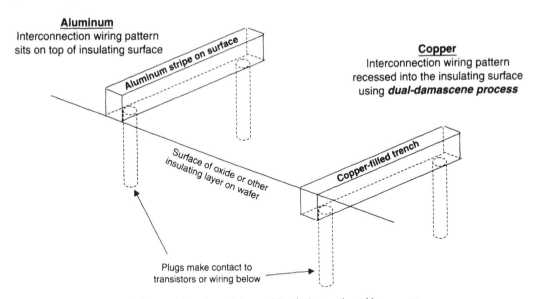

- Aluminum wiring still widely used but being replaced by copper

Figure 6-14: Copper Damascene

The photolithography process is straightforward. The challenges for photo at this level are alignment of the pattern and creating a mask to etch thin trenches. The trenches must be positioned precisely over top of the tungsten plugs, a job made harder because of the narrowness of the trenches.

The etch presents many challenges, as well. If the low-k material is an inorganic film, then the etch is similar to a standard oxide etch. Organic low-k materials can be much more difficult to process and require a hard mask because the organic low-k is chemically similar to the resist; etching it will also etch the resist mask at a rather rapid rate. Other types of innovative films require highly specialized chemistry and tools.

Because of the differences in low-k materials, there is no standard process that has been adopted by the industry as yet. This is one of the most exciting examples of R&D going on in the industry today and will produce many fascinating advances as the work proceeds.

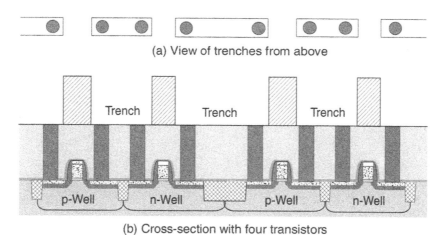

(a) View of trenches from above

(b) Cross-section with four transistors

Figure 6-15: Completed Trenches

Copper First Level Interconnection Process

7.1 Ta/TaN Barrier Layer Deposition

Copper is a very dangerous contaminant in silicon devices. How can it be used safely in making a chip?

The story of copper's entry into the chipmaking process is one of acute paranoia. Copper, other common metals, sodium, and in fact, a large portion of the periodic table will contaminate the chips and kill the devices. Great care in manufacturing is required.

The motivation to integrate copper is very high. Copper is a better conductor than aluminum and, incidentally, has a low rate of electromigration. The chips will certainly be faster and more reliable with copper, right?

Only if the copper can be kept away from the transistors! It is still the practice in many fabs to complete the chips up to the point where they are ready for copper deposition and then the wafers leave the "front end of the [manufacturing] line," never to return. The copper must be kept away from the wafers in the early stages of the process to ensure that they are not contaminated.

Terminology

Acronyms abound in the semiconductor industry. FEOL stands for "front end of the line" and BEOL stands for "back end of the line." FEOL is defined as the processing steps that begin with the starting silicon wafer through and including contact etch. All of the processing steps after contact etch, until the wafer leaves the fab are done in the BEOL.

This terminology emphasizes the difference in the processing steps involving metal deposition and all those prior to the introduction of metal on the

---

wafer. Many metals, particularly copper, are dangerous contaminants and cannot be allowed to contact unprotected surfaces in the layers containing the transistors.

Barrier layers must completely seal in the copper to prevent it from migrating in any direction. For the walls of the copper-filled holes and trenches, a barrier layer is deposited before the copper is deposited. Later, a barrier layer is needed to cover the top of the copper-filled trenches; the process is discussed in Chapter 7. Currently, the most widely used barrier layer for holes and trenches is a tantalum/tantalum nitride combination of films. Ta does not react with Cu and forms strong metal-metal bonds so it is a good glue layer. TaN blocks the diffusion of contaminants so it is a good barrier material. The two films form an almost ideal composite.

Ta and TaN are deposited using PVD technology. Development work is showing promise for TaN deposition using a new technology called *atomic layer deposition (ALD)*. It produces a film with excellent conformality, allowing it to be as thin as possible with much lower resistance to current flow. ALD is a method of assembling films a few atoms at a time into layers whose thickness may be specified in numbers of atomic layers. These films are of excellent integrity and structure. ALD will be an interesting topic for future discussion as it moves into production.

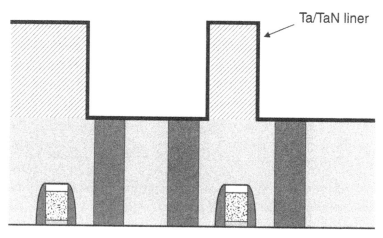

Copper must be encapsulated to prevent out-diffusion
Figure 6-16: Deposited Barrier Layer

7.2 Sputter Copper (Cu)

Copper trench fill is a two-step process. The optimum way of depositing the bulk of the material is electroplating. But that presents a problem. Electroplating requires an unbroken film of copper on the wafer; a conducting coating is required for plating to occur. So a seed layer is needed.

The seed layer is deposited on the wafer using the familiar PVD method. Careful sputtering will coat all of the surfaces with a thin Cu film, including the interiors of the small trenches.

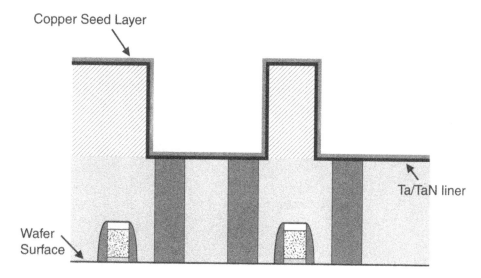

Figure 6-17: Cu Seed Layer

7.3 Electroplate Copper (Cu)

Electroplating was found to work better than sputtering to completely fill the small features on the wafer. The wafer with the conducting seed layer acts as the cathode in the electroplating bath. Copper plates out on the seed layer covering all the surfaces of the wafer.

Electroplating has been used for many years by the printed circuit board industry to form metal wires and interconnects. The technique is so well developed, it could be introduced to the fab for copper deposition at low cost with high throughput. Figure 6-18 illustrates the operation of the plating bath. In practice, the wafer is held horizontally and the solution moves past it for better overall deposition characteristics.

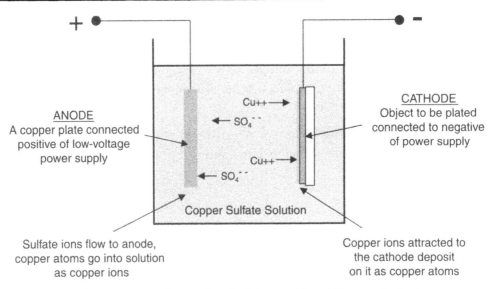

Figure 6-18: Electrochemical Deposition (Electroplating)

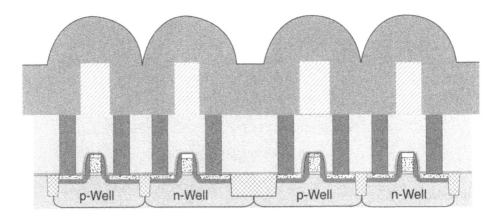

Figure 6-19: Copper Deposited to Fill Trenches. Copper also plates on the top surfaces.

7.4 Copper CMP

The excess copper is polished off in much the same way as the tungsten. This metal polish is somewhat different from the W because the material characteristics are different. Cu polishing is perhaps the most demanding CMP process in use today.

The polish must remove the conductive Ta/TaN layer under the copper and stop on the IMD. Considering that it can be a relatively soft or mechanically weak low-k material, other problems may appear for the CMP engineer in preserving the integrity of the film.

Copper connections
between tungsten plugs

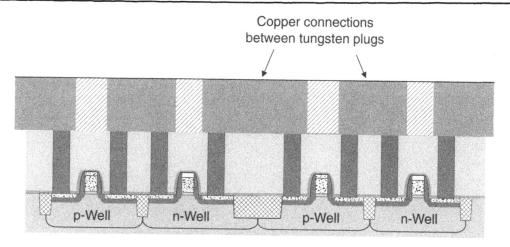

Figure 6-20: Completed First Level Metal

The wafers may now proceed to the next major operation in the manufacturing process: multilevel metallization.

CHAPTER

7

Multilevel Metal Interconnects and Dual Damascene

Introduction

In this chapter, you will learn:

- How multilevel interconnects are made
- How the dual damascene process works
- How the finished chip is protected from the environment

Multi-Level Metallization Process Flow

1. Deposit the Silicon Carbide (SiC) Barrier Layer
2. Deposit low-k IMD
3. Dual Damascene Process
   a. Mask and etch vias (hard mask process)
   b. Remask and etch trenches
   c. Deposit Ta/TaN barrier layers
   d. Sputter copper seed layer
   e. Electroplate copper
   f. CMP to remove excess copper
   g. Build additional layers
      i. Deposit SiC barrier layer
      ii. Repeat dual damascene process
4. Form Bonding Pads
   a. Metal deposition
   b. Photo/etch for bonding pads

5. Final Passivation Process
    a. Deposit final passivation
    b. Photo/etch for bonding pads

Key Points in the Multi-Level Metallization Process

1. The damascene process is necessary for copper technology

2. Copper requires the use of effective barrier films

3. Low-k dielectric materials are replacing deposited oxide as the intermetal dielectric

4. A hard, impermeable coating must cover the finished chip for protection from contaminants

Each of these points will be explained in the appropriate sections of this chapter.

Discussion

Making all of the required electrical connections on the chip cannot be done with only one layer of metal "wires." With hundreds of millions of transistors and countless other components on the chip, the wiring has become very complex. It is common for eight to ten crisscrossing layers of metal to be necessary for state-of-the-art chips.

Fortunately, each layer is constructed in essentially the same way, so this chapter will suffice in describing that process. Bear in mind that the metal wires get thicker in the upper interconnection levels because they need to carry more current.

Originally, there were two approaches to choose from in constructing the dual damascene structure. One approach used a silicon nitride layer positioned in the center of the IMD. That layer was called the *etch stop*, and was used to ensure that the trench etch step did not penetrate too far into the IMD and that the trench had a flat bottom. The trouble with that design was the high-k value of the nitride; it raised the overall k-value of the IMD stack, reducing the effectiveness of the low-k IMD. As a result, the nitride etch stop is not often used.

Control of the trench depth is handled with a well-designed etch process. The depth is important because it affects the thickness, and therefore the resistance, of the copper wire that is formed within the trench. Also, the trench must be kept a minimum distance from the underlying conductors in order to minimize electrical interaction. The chip designer will specify the dimensional tolerance used by manufacturing (see Chapter 2).

The final passivation of the chip will also be discussed in this chapter. At the end of the manufacturing process, the chip must be protected from environmental contamination before it can leave the cleanroom. The wafer is coated with a tough covering that supplies the needed protection. But if the chip is covered in a tough protective film, how are the electrical connections attached to it? The answer to that question is found in Section 5.

Deposit Barrier Layer and Intermetal Dielectric

After the copper CMP step discussed at the end of Chapter 6, bare copper is exposed at the top of the trenches. To prevent copper from migrating into the low-k dielectric layer that will be deposited on top of it, a diffusion barrier is needed.

PECVD silicon nitride has been used for many years as the barrier layer at this point in the process. However, it is an important issue to keep the overall k-value for the film stack as low as possible. The k-value for SiN is about 8 for this application. So a change is often made in barrier layer materials.

A familiar example of a material that is often used to replace silicon nitride is silicon carbide (SiC). SiC is usually deposited using a CVD process. This film is much like some of the other films discussed earlier in the text in that its composition may not be very close to conventional SiC produced in higher temperature reactions. Nevertheless, it works well as a barrier layer and can be deposited at the required lower temperature.

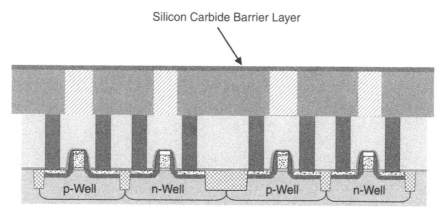

Figure 7-1: SiC Barrier Layer

SiC has a lower k than SiN, between 5 and 6, so it improves the performance of the chip. In addition, it is at least as good a diffusion barrier as SiN. It sticks well to many IMD's and to copper so it will not cause the film stack to delaminate.

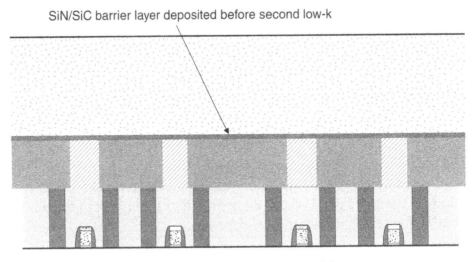

SiN/SiC barrier layer deposited before second low-k

Figure 7-2: Second IMD Deposition

The same low-k material is often used for the second and subsequent IMDs. But sometimes it is advantageous to use a different material.

See Chapter 6 for a discussion of this deposition and low-k IMDs.

Dual Damascene Process

3.1 Introduction

Definition: *Dual Damascene (DD)* is the name of a process for making metal interconnections that is reminiscent of the metal inlay techniques used in the Middle
East since the middle ages. The name originates in Damascus, the capital of
modern Syria.

Purpose: The dual damascene process is needed for copper technology. Very fine
metal wires cannot be etched in copper as they were in aluminum because there is
no practical etch process available for copper. A major constraint is the volatility
of the reaction products. If the chemical reaction in the etcher produces chemicals
that coat the wafer and bind to the surface rather than evaporating off the surface,
then new reactant is blocked from reaching the film to be etched. The etch stops
without transferring the desired pattern. For example, chlorine gas is often used
to etch metals in plasma etchers; it will attack the copper but the resulting copper
chloride product of reaction will not readily evaporate.

Discussion

The same approach is used for dual damascene as that discussed in Chapter 6 for
single damascene: a trench is filled with copper and the excess copper is polished
off with CMP. But in the single damascene example the contact holes were already
filled with tungsten. How is the connection to the lower metal layer made with dual
damascene?

The origin of the name "dual" damascene is the answer. Two process steps must
be performed, one to make holes in the IMD (the vias) so that the copper connections

to the underlying metal layer can be made, and a second step to make the trench that will be filled with copper. The next section begins the discussion of this sequence of steps.

The two photo/etch steps in the dual damascene process can be performed in any order: trench-first then via or via-first then trench. However, the via-first process is preferred because of patterning difficulties in the trench first method. Other creative sequences of steps have also been tried without much success. The via-first process will be discussed here.

3.2 Via Photo/Etch

Definition: A *Via* is a hole made in the IMD similar to the contact hole discussed in Chapter 6. The name was taken from the printed circuit board industry where a similar structure is used on PC boards.

Purpose: The via will be filled with metal in order to make electrical connection to the underlying metal interconnect layer. The hole is not unlike the contact hole except that the etch stops on copper rather than on silicon.

Discussion

Before discussing the particulars of via photo and etch, it should be noted that if the IMD is an organic material or any other vulnerable film then a hard mask must be used. As explained in Chapter 6, the photoresist and organic IMDs are very similar in chemical properties and will etch at similar rates in the same etch environment. That would damage or destroy the mask before the pattern was correctly reproduced in the film.

Should a hard mask be needed, a deposition step would precede the photo/etch sequence. Typically, silicon nitride would be used as the masking material. Then an etch step to make the hard mask would precede the IMD etch.

The challenge of the via-first DD process is that the via holes must be etched through a very thick IMD film. Not only does the thickness of the IMD underneath the copper interconnects need to be large enough to sufficiently insulate the stacked layers of conductor; the thickness of the metal contained in the trench is added to the IMD thickness. The result is a very high aspect ratio hole.

The photo/etch process is much like the contact recipe for inorganic IMD materials. Organics require much different etch recipes and will often include oxygen as one of the components. The hard mask material used for the organics is often SiN, or an oxide so the familiar fluorine chemistry will produce the hard mask.

The etch chemistry will differ somewhat from contact etch because of the different underlying material and the (sometimes) higher aspect ratio. In addition, there are so many choices for low-k IMDs that the etch chemistry can potentially be very different.

It must not be overlooked that the SiC barrier layer covering the copper at the bottom of the via must also be removed. The electrical connection between copper layers would be poor or nonexistent if this film were left in place.

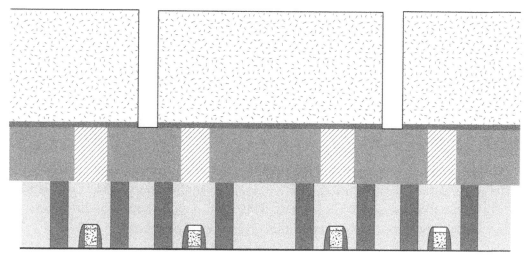

Figure 7-3: HAR via Cross Section

3.3 Trench Photo/Etch

This operation is a duplicate of the one described in Chapter 6. The wafer returns to photo, is again coated with resist, exposed and developed to produce the trench pattern.

Notice that the vias are present on the wafer. It is deliberate that they fill partially with photoresist. The resist "plug" protects the underlying film from possible erosion or damage during the trench etch. The resist in the vias is removed during the post-etch clean.

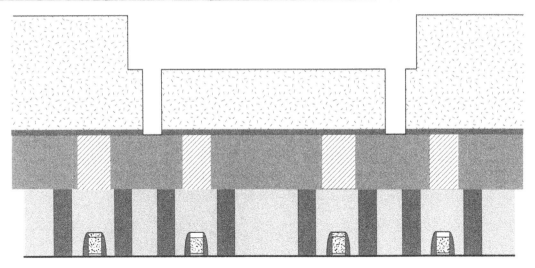

Figure 7-4: DD via Trench Formation

3.4 Deposit Barrier Layers

The tantalum and tantalum nitride barrier stack is the standard for lining the vias and trenches.

At this point it can be seen that the trenches and vias are completely coated with barrier materials. Note that these layers are depicted in the enlarged view of Figure 7-7. Ta/TaN will separate the Cu from the IMD. SiC covers the top of the underlying copper and will be deposited on top of this layer of copper shortly.

3.5 Sputter Copper

The seed layer of Cu is sputtered exactly as described in Chapter 6.

3.6 Electroplate Copper

The same electroplating process described in Chapter 6 is used here.

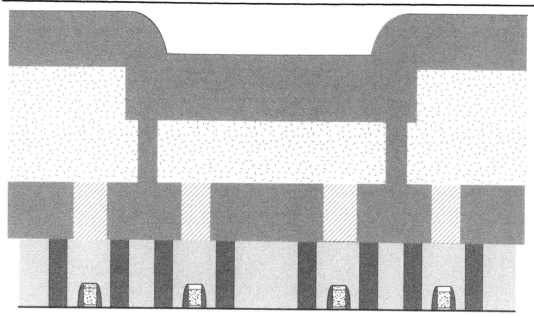

Figure 7-5: Bulk Copper Deposition. Additional detail shown in Figure 7-7.

3.7 CMP to Remove Excess Copper

This copper polish is the same as performed in Chapter 6. The polish stops on the IMD. Remember that the Ta/TaN barrier layer under the copper must be polished off the top of the IMD so that all of the Cu interconnects are not shorted together.

Figure 7-6: Completed DD. Additional detail shown in Figure 7-7.

3.8 Deposit SiC Barrier Layer

The top seal on the copper is done with silicon carbide as before.

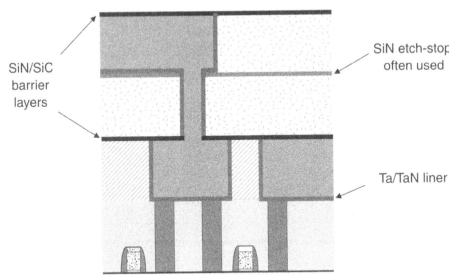

SiN/SiC barrier layers

SiN etch-stop often used

Ta/TaN liner

Copper must be encapsulated to prevent out-diffusion

Figure 7-7: Completed DD with SiC Barrier Layer

3.9 Build Additional Layers

The creation of multiple metal layers is simply a repeat of the previous steps.

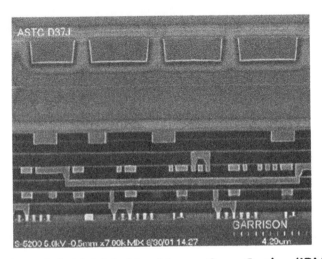

Figure 7-8: Multiple Metal Layers Cross Section (IBM)

Form Bonding Pads

Definition: *Bonding Pads* are large pads of metal that are the end connection of all the chip circuitry.

Purpose: The metal wires from the chip package are attached to the bonding pads. The pads are typically 90×90 microns, so they are very large compared to most of the other features created on the chip.

Discussion

The bonding pads are generally made of aluminum or copper, and sometimes, aluminum deposited on top of copper. Bonding to the aluminum pads, using the standard gold or aluminum wires from the chip package, works very well. Wire bonding is discussed in the next chapter.

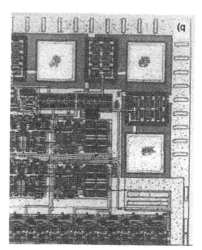

Figure 7-9: Typical Bonding Pads *(Semiconductor Picture Dictionary)*

Final Passivation Process

5.1 Deposit Final Passivation

The final protective passivation film is commonly a stack of two films. A layer of PSG is put down first, followed by silicon nitride. Historically, either of those films have been used independently but the trend has been to combine the strengths of both by simply stacking them up, nitride on top of PSG.

The PSG film acts something like the pad oxide seen in Chapter 4. The expansion and contraction of the chip cannot be allowed to crack the hard nitride layer, so the oxide relieves the stress.

SiN is deposited on top of the PSG. It is a good scratch protector as well as a tough barrier to contaminants like sodium and other alkali ions.

Silicon oxynitride is sometimes used as the final passivant. It is a film with a flexible composition, somewhere between oxide and nitride.

5.2 Photo/Etch for Bonding Pads

The bonding pad pattern is especially important because the passivating film must only be removed from the top-central portion of the pad while excluding the pad edges. The passivation must overlap the edges of the pads in order to maintain the protective seal that protects the chip. Environmental contaminants will drastically reduce the useful life of a chip.

This center section illustrates the overlap of the pads that must be maintained.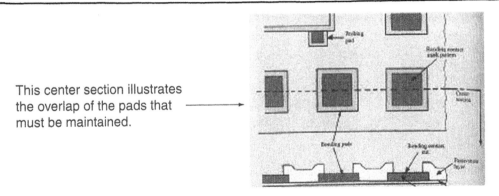

Figure 7-10: Bonding Pad Openings
(Semiconductor Picture Dictionary)

The wafers are ready for the last sequence of steps leading to completion: Test & Assembly.

CHAPTER

8

Test and Assembly

Introduction

In this chapter, you will learn:

- The ways in which wafers and chips are tested

- How the chips are packaged

- Why different types of packaging material are used

Chapter 1 introduced the "back-end" processing topics that are expanded in this chapter. Testing and packaging finish off the job of making the end product, a reliable, useful electronic component.

Test and Assembly is a very large segment of the industry but a detailed discussion is beyond the scope of this book. The reader is invited to enjoy a more thorough treatment of this topic in several of the works referenced in the Bibliography.

Figure 8-1: Packaged Chips. Lid removed from large pin grid array (Intel Pentium) to show die inside.

Terminology Reminder

Here is another instance of terminology that may be confusing because of a lack of consistency from fab to fab. As mentioned in Chapter 7, the term "back end of line," or BEOL, is sometimes used to refer to the fab operations that begin with first metal. That distinction is becoming more common today since the use of copper has forced such a distinct break in processing procedures when it comes time to deposit the copper. However, the test and assembly part of chipmaking has been called the back-end for most of its history, so that terminology will continue to be used here.

Another terminology confusion results from the differing tests that are done to the large variety of device types manufactured in the industry. Some manufacturing lines will lump operations together, at least as far as terminology goes, if not in actual procedures. An example of a term with variable meaning is "e-test," which will be discussed shortly. E-test is often a nickname for "electrical test." Some companies perform the operation that they call e-test as part of wafer sort (probe). Some use the term to reference only parametric testing; still others use it to describe both parametric test and wafer sort. Some companies do not use the term at all. So be prepared for the need for clarification.

This discussion will use the standardized textbook terminology.

Why perform the extensive testing that is described in the next section? In light of all the rigorous checks and sophisticated procedures employed in the manufacturing process, is chip reliability and longevity an issue?

The answer to that question is an emphatic "yes!" It is hoped that the reader has, by now, some appreciation of the difficulty of making these devices with their unimaginably tiny components. The probability for imperfections is high. Considering the critical functions performed today by many chips, unreliable products would too often risk lives. In addition, computers are indispensable to financial institutions, commerce and industry; the cost to businesses of poor quality chips would be enormous.

Military applications of semiconductor devices expand the necessity of reliability. The so-called mil-spec requires electronic components purchased by the military to tolerate the most extreme conditions imaginable. Some chips must continue to function even after exposure to the electromagnetic burst radiation from an atomic explosion.

The testing process is important for another reason. It provides critical feedback information to the manufacturing engineers and the design engineers about the quality of their work. Many details about the process of designing and making chips are improved based upon test data. Indeed, an idea may look good on paper and seem to be easy to manufacturer, but if the chips don't work, it's all for naught.

Wafer and Chip Testing

Wafer and Die Tests

1. In-line Parametric

2. Wafer Sort

3. Final Functional

2.1 In-line Parametric Test

Each stage in the manufacturing process includes measurements and tests. This monitoring is done using product wafers, and also through the use of specially prepared monitors called by various names such as pilot wafers, test wafers or monitor wafers. These checks must be done to ensure that the design specifications are correctly followed and that the manufacturing tools are performing properly.

The term "parametric" refers to the "parameters" that are measured during the testing process. Parameters are the measurable or quantifiable characteristics that are specified by the design engineers.

In-line testing may include a vitally important set of electrical measurements, (often called *e-test*), that are made before the wafers are sent on to wafer probe. Dozens of basic measurements are made on individual transistors and other test structures in test patterns designed into the "scribe lines," the spaces between the chips.

Some simple examples of the tests made are:

1. Gate oxide breakdown voltage

2. pn junction breakdown voltages

3. Transistor threshold voltages

4. Contact resistances, metal to silicon

5. Gate line resistance

The results of this test are very important in ensuring that the wafers are being processed correctly, and in providing feedback to the fab engineers of the need for any adjustments. If the parameters measured at this point on a wafer do not meet specifications, that wafer will be rejected to avoid the substantial time and cost of the functional test in wafer sort. Such wafers may receive extensive failure analysis to find the source of the problems.

2.2 Wafer Sort (Probe)

Wafer sort, or *probe*, is a functional test that is done to every die on the wafer at the end of the manufacturing process, while the wafer is still intact. Each die is tested with a probe that has many tiny electrodes. A computerized system performs the tests on the chip and collects statistical data on the product.

The test system keeps a map of passed and failed die on every wafer. Failed die are identified in a computer map, or marked with an ink dot; they are said to be "inked out." After the wafers are separated into individual chips, a special tool lifts off the good die and places them in chip packages. Failed die are ignored by the tool.

Figure 8-2: Probe Card or Wafer Being Probed

A familiar term related to probe is "sort-yield." Sort-yield is simply the ratio of the number of functional die to the total possible number of die. This yield figure has enormous impact on a company because it directly evaluates the effectiveness of its manufacturing line. It is possible for a mature product coming from a top quality manufacturing line to see sort-yields in excess of 95%. However, some products are simply too difficult to manufacture and never get anywhere near that figure.

Historical Change in Sort-Yield

As recently as twenty years ago, it was possible for a new product introduced to the manufacturing line (first silicon) to have a sort-yield of zero (0) in the first manufacturing attempt. Of course, it didn't always happen, but it was still expensive and risky to introduce new products in those days.

Times are somewhat better today. It is common to see 15% or better sort yield for first silicon. Manufacturing practices have become standardized, yield management procedures established and design rules adjusted to match the ability of manufacturing to produce a working finished product. However, new, untested technology carries the same risk as ever and certainly costs a lot more. It may take three to five years of development before test chips indicate that it is possible to attempt first silicon.

2.3 Final Functional Test

Final functional testing is accomplished after the chips are packaged. The next section discusses the process of changing the wafers into individual electronic component parts. It will be clear from that discussion that several additional potential sources of damage are added during assembly and packaging. In addition, some tests cannot be done until the chip is packaged, notably high speed testing. So final functional testing is essential to ensure that the product that is shipped to the customer is indeed functional and reliable.

Some types of electronic components are both electrically and thermally stressed before testing. This process is called *burn-in*, and is used to eliminate marginal parts that made it up to this point. The chips are plugged into special sockets that mount in ovens. The electrical and thermal cycling can go on for days or weeks. This operation forces any weak chips to fail. Of course, not all types of chips require burn-in.

The final testing regimen is designed to ensure that the finished product meets the product specifications. It will also weed out any chips that were damaged during the packaging operation.

Many chips are designed with redundancy in the memory arrays. This clever scheme allows the manufacturer to blow fuses at final test to remove damaged or imperfect sections from the circuit; the fuse reroutes the circuit and bypasses the bad sections. The chip is still fully functional and can be sold for a good price. However, the highest price still goes to the perfect chip that tests out with the fastest speed. The categorizing of the chips is a process called *binning*. There is a "bin," or category, for perfect parts and another for repaired parts; even nonfunctional parts have a bin depending on the type of failure found during testing. The final test process assigns all of the chips to the appropriate bin.

Assembly and Packaging

Back-End Process Flow

1. Die separation

2. Die attach and bond pad connection

3. Encapsulation

4. Electrical testing

5. Pack and ship

3.1 Die Separation

After the wafers are probed and leave the fab, another operation is commonly performed that grinds the backs of the wafers called, not surprisingly, *back-grinding* or *back-lapping*. It thins the silicon substrate, which helps to keep the final package as small as possible and is often important in maximizing heat conduction out of the chip.

Following back-grinding, the wafers are mounted on special adhesive sheets of plastic held in a frame. The frame fits into a special machine that will saw the individual die (chips) apart. Recall that the chips are separated by a space called the *scribe line* or *street*. The die separation is dimensionally matched to the saw (dicing tool) so the saw blade will not damage the parts.

The dicing saw uses a diamond-impregnated blade that spins at about 20,000 rpm. The entire process requires precise alignment of the saw blade and involves many other steps. The blade penetrates the wafer but does not cut through the plastic. When the dicing is completed, the adhesive on the plastic holds all of the individual die in place to make it easier to put them into their individual packages.

Figure 8-3: Dicing Saw (SEMI)

3.2 Die Attach and Bond Pad Connection

The die are ready to be mounted on the substrate portion of the package. This operation is called *die-attach* or *die-bonding*. A special machine lifts each die from the plastic using a vacuum handling arm and places it on the package substrate (pick-and-place). This is the machine mentioned earlier that ignores the bad die. The chip is attached to the package which has been prepared with adhesive, often epoxy, but sometimes other attachments like a special solder of gold-germanium alloy is used.

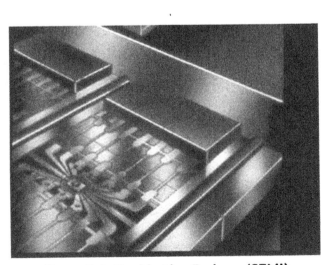

Figure 8-4: Dual-in-Line Package (SEMI)

The dual-in-line package is the most common package type, and contains a lead frame to which the chip is attached. The lead frame is a mounting that has the appropriate number and configuration of wires, called "leads," to attach to other electrical devices, circuit boards and the like.

Making the electrical connections between the bonding pads on the chip and the leads on the package is done in several different ways. The most familiar is the wire bonding method, where tiny gold or aluminum wires are attached to the bonding pads and to connection points on the leads. Recall that the bonding pads are very small regardless of the fact that they are the largest features that are found on the chip.

The gold wire is attached using thermocompression, ultrasonic or thermosonic techniques. Aluminum is attached with the ultrasonic technique. For example, the gold wire is threaded through a tungsten carbide capillary. The wire extending past the end of the capillary is exposed to a hydrogen flame and melted into a ball. The ball is then pressed onto the bond pad while ultrasonically exciting both the wire and the pad, heating both to about 150°C and forming the bond (thermocompression bonding). The capillary then loops the wire to the attachment point on the lead. The wire is pressed against the contact point while exciting the wire with ultrasonics, forming a wedge bond. The wire is easily broken off at the wedge bond and the capillary moves to the next pad to continue the process.

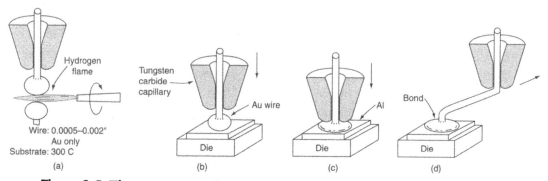

Figure 8-5: Thermocompression Ball/Wedge Bonding of Gold Wire (Ghandhi)

Wire bonding is fully automated and the speed of the tool is incredible. The connections are done faster than the eye can see.

Some chip types are configured such that wire bonding is impractical or impossible. There is also the case of certain high pin count chips that make wire bonding difficult. One alternative is the tape automated bonding (TAB) method. TAB requires the fabrication of solder bumps on the bond pads, a process requiring several steps, that is done before dicing. In addition, the tape undergoes fabrication to produce the metal lines (wires) that match the bond pad pattern on the chip. Then, prior to die attach, the chip is bonded to the tape using thermocompression. Next, instead of using a separate die-attach step, the entire chip is encapsulated. Actually, a large group of chips are encapsulated while still on a continuous strip of tape. After packaging, they are cut apart.

Flip-chip bonding is another method that requires the production of solder bumps. The solder bumps used for this application are made of a metal alloy with a melting point that best fits the process requirements. The entire chip can be covered with bonding pads rather than the restriction of some types of processes that only allow pads along the edges of the chip. The device is soldered directly to the package substrate.

3.3 Encapsulation

Encapsulation is the process of placing chips in protective packages. Many packages are available for chips. The intended use of the part is key in deciding how to encapsulate it. Military hardware will be exposed to harsh environments so tough packaging materials are used. Simple plastic packages, such as the dual in-line package, suffice for many chips. Metal, ceramic, plastic and composite packaging materials are all widely used.

The packaging material is matched to the chip function. If necessary, the package dissipates excess heat to some extent but some packages must be placed in contact with an additional large heat sink; some microprocessors even need a built-in fan! The package must protect the chip from moisture, dust and other environmental contaminants. Some packages are designed to block radiation of certain types, and the package can affect the electrical properties of the chip, so compatibility is necessary.

Encapsulation is a process that varies enormously with the package type. Ceramic packages are processed quite differently from plastic packages, which are different from composite packages. The reader is invited to refer to the bibliography for suggestions for further reading on these processes.

Now that final functional testing and packaging is complete, the packages are labeled. The parts are then packed in shipping containers and sent to stores or customers. The entire chipmaking process is now complete.

APPENDIX

A

Science Overview

Introduction

The semiconductor manufacturing industry is interdisciplinary in the extreme. Principles of physics, chemistry, materials science, electrical engineering, mechanical engineering and other disciplines are all an integrated part of the job of making chips. During the many years that we (the authors) have taught seminars to semiconductor professionals, we have learned that the scientific background knowledge of our students has varied widely. It has been a challenge to offer courses that are appealing to the general audience while maintaining a quality of content that expands the student's horizons. It has been essential to include a science overview to put everyone on the same page, so to speak, where science vocabulary and science principles are concerned.

In this appendix, you will find all of the essential scientific concepts that are needed to understand the discussion found in the text. Of course, many readers are not in need of all the sections of this tutorial. Feel free to review only the material that is new and informative to you.

This appendix includes relevant discussions of:

- Atomic Theory: atoms, molecules, ions, states of matter
- Chemical principles for semiconductor manufacturing
- Some characteristics of solids
- Some useful concepts about electricity

Integrated circuits are very small. In fact, the component parts of these electrical devices are almost as small as it is possible to make them and still use existing technology, which deals with bulk materials.

With these facts in mind, it seems like a helpful idea for us to begin our discussion of chipmaking with a review of some of the principles that make it all possible.

Atoms and Molecules

1.1 The Atom

What Is An Atom?

An atom is the smallest unit of an element. It is made up of electrons that surround a nucleus made of protons and neutrons, except for hydrogen, the only atom with no neutrons. The radius of a typical atom is about one Angstrom unit (one ten-billionth of a meter).

The Oxford American Dictionary says that an atom is "the smallest particle of a chemical element that can take part in a chemical reaction." However, not all atoms will take part in a chemical reaction, so perhaps it is possible to be a bit more explicit. One freshman chemistry text (Brown & LeMay, see Bibliography) defines an atom as "the smallest unit of an element that can combine with other elements." Again, that definition leaves out some necessary information. It is true that most atoms will combine to form compounds and molecules, but not all of them will do so naturally.

The real secret of the nature of the atom is in the number of protons in its nucleus. Any change in the number of protons and the chemical characteristics of the material become very different. That fact helps us write a more generalized definition of an atom.

Definition: An *atom* is the smallest unit of an element that still has all of the same chemical characteristics of the element.

Any attempt to make an atom into a smaller bit of matter would change the chemical properties because it would require removing nuclear particles (protons and neutrons). As we shall see later, some electrons can be removed or added (making an ion) or shared in chemical bonds without altering the basic chemical characteristics of the atom.

If you divided a piece of an element like aluminum (Al) into smaller and smaller bits, eventually you would produce the smallest bit of aluminum that is still chemically identifiable as aluminum, an atom of Al.

The nucleus of the atom contains the protons (positive charge) and neutrons (no charge) except for hydrogen, the oddball, which has only one proton in its nucleus. Electrons (negative charge) surround the nucleus and are most easily pictured as if they were planets orbiting the nucleus, something like the way that planets orbit the sun in our solar system.

It is interesting to note that the electron is very, very small. It is more than 1,830 times lighter than a proton. Protons and neutrons weigh almost exactly the same amount.

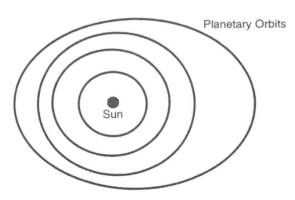

Illustrates one model of
how electrons orbit the nucleus

Figure A-1a: Simplified Solar System

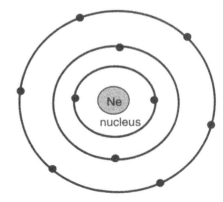

**Figure A-1b: Filled Electron
Orbitals for Neon**

I   VIII

I	II											III	IV	V	VI	VII	VIII
¹H Hydrogen	II											III	IV	V	VI	VII	²He Helium
³Li Lithium	⁴Be Beryllium											⁵B Boron	⁶C Carbon	⁷N Nitrogen	⁸O Oxygen	⁹F Fluorine	¹⁰Ne Neon
¹¹Na Sodium	¹²Mg Magnesium											¹³Al Aluminum	¹⁴Si Silicon	¹⁵P Phosphorus	¹⁶S Sulphur	¹⁷Cl Chlorine	¹⁸Ar Argon
¹⁹K Potassium	²⁰Ca Calcium	²¹Sc Scandium	²²Ti Titanium	²³V Vanadium	²⁴Cr Chromium	²⁵Mn Manganese	²⁶Fe Iron	²⁷Co Cobalt	²⁸Ni Nickel	²⁹Cu Copper	³⁰Zn Zinc	³¹Ga Gallium	³²Ge Germanium	³³As Arsenic	³⁴Se Selenium	³⁵Br Bromine	³⁶Kr Krypton
³⁷Rb Rubidium	³⁸Sr Strontium	³⁹Y Yttrium	⁴⁰Zr Zirconium	⁴¹Nb Niobium	⁴²Mo Molybdenum	⁴³Tc Technetium	⁴⁴Ru Ruthenium	⁴⁵Rh Rhodium	⁴⁶Pd Palladium	⁴⁷Ag Silver	⁴⁸Cd Cadmium	⁴⁹In Indium	⁵⁰Sn Tin	⁵¹Sb Antimony	⁵²Te Tellurium	⁵³I Iodine	⁵⁴Xe Xenon
⁵⁵Cs Cesium	⁵⁶Ba Barium	⁵⁷La* Lanthanum	⁷²Hf Hafnium	⁷³Ta Tantalum	⁷⁴W Tungsten	⁷⁵Re Rhenium	⁷⁶Os Osmium	⁷⁷Ir Iridium	⁷⁸Pt Platinum	⁷⁹Au Gold	⁸⁰Hg Mercury	⁸¹Tl Thallium	⁸²Pb Lead	⁸³Bi Bismuth	⁸⁴Po Polonium	⁸⁵At Astatine	⁸⁶Rn Radon
⁸⁷Fr Francium	⁸⁸Ra Radium	⁸⁹Ac* Actinium															

Figure A-2: Periodic Table of the Elements (Abbreviated)

Here is an abbreviated version of the Periodic Table of the Elements. This abbreviated table includes most of the naturally occurring elements and all of the elements that we will be discussing in this book.

Several groups of elements will be discussed in this book. The columns that make up the Periodic Table contain families of chemicals, the elements with similar chemical properties. These families are called *Groups*. Several groups are of particular interest to this discussion.

Group VIII is called the *noble gases*. This family of chemicals is essentially nonreactive. Also called the *inert gases*, these elements do not participate in any chemical reactions that are of interest to us in the semiconductor industry; they are still helpful in creating the structures on the wafer because of their physical properties so we use some of these gases extensively.

Group VII, called the *halogens*, is very different from Group VIII. This family of chemicals includes some of the most reactive substances in the universe. The high degree of reactivity is a great help in making chips and several of these chemicals are used in chip fabrication.

Groups III and V are interesting because some semiconductor materials are produced in chemical combinations of elements from these two families. These are, not surprisingly, referred to as *III–V semiconductors*.

Of course, Group IV is very interesting because it contains silicon, the most familiar semiconductor as well as germanium, the element used to make the first transistor and the first generation of semiconductor products.

We will be pointing out the properties of many elements from the Periodic Table throughout our discussions. It is very interesting to note that elements farther to the left side of the table are often highly damaging contaminants that can ruin the devices that we are making, even in trace amounts.

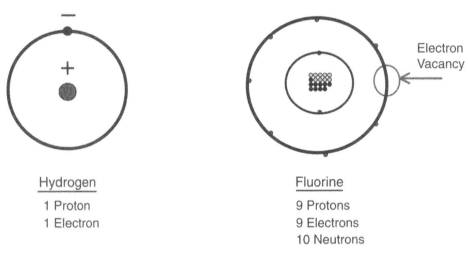

Hydrogen

1 Proton
1 Electron

Fluorine

9 Protons
9 Electrons
10 Neutrons

Electron Vacancy

Figure A-3: H Atom and F Atom

1.2 Molecules

Definition: A *molecule* is made up of atoms that are bound together. The previously described definition for atoms applies to molecules with only one small change: "A molecule is the smallest fundamental unit of a *chemical compound* that still retains all of the chemical characteristics of the chemical compound." While some compounds are very nonreactive, they are nevertheless composed of atoms so we won't split hairs this time.

Definition: A *chemical compound* is made up of two or more atoms that are bound together by chemical bonds. A compound made up of a variety of elements will have distinctly different chemical properties than the separate elements of which it is composed.

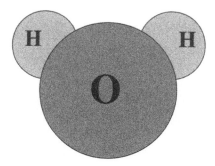

Figure A-4: H$_2$O Molecule

1.3 Organic Molecules

Organic molecules are defined as chemical compounds in which carbon atoms are a principle component. There are a very few carbon compounds that are not organic, such as carbon dioxide and silicon carbide. Suffice it to say that there are other rules that define organic compounds that are beyond the scope of this discussion.

Why call them "organic" you may ask? All living things are primarily composed of these special carbon compounds. The interesting thing about some organic molecules is in the way they tend to form long chains called *polymers*. These chains are made of many multiples of a basic building block called a *monomer*. Figures A-5 and A-6 show the monomer and polymer of two familiar "plastics" that you have around the house (plastics are made of organic molecules, too).

```
Monomer: Ethylene          Polymer: Polyethylene

   H   H                    H   H   H   H H
   |   |                    |   |   |   | |
   C = C                  - C - C - C - C - C -
   |   |                    |   |   |   | |
   H   H                    H   H   H   H H
```

Figure A-5: Ethylene and Polyethylene

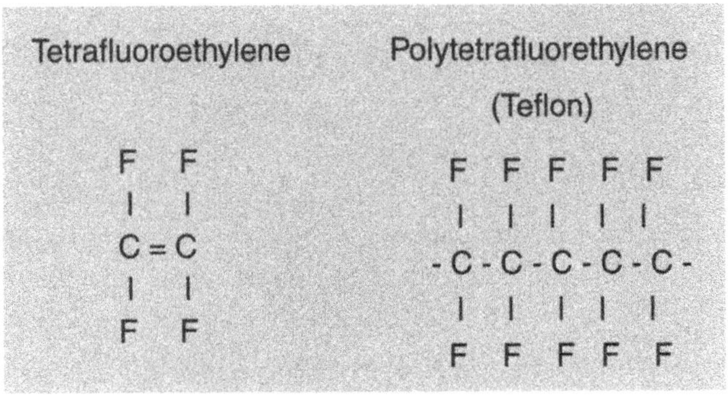

Figure A-6: Tetrafluoroethylene (Teflon®)

States of Matter

Matter comes in several forms called *states*. Everyone is familiar with the three most common states: solid, liquid and gas. There are two others, the plasma state and the Bose-Einstein Condensate. Plasma will be discussed in this section. The Bose-Einstein Condensate is really rare and hard to produce, putting it beyond the scope of this text.

If you reference the Periodic Table (you may need to look at a complete version for this, just to be sure), and look for elements that occur naturally or can be refined and remain in their elemental form at room temperature in air for some period of time, you will see an amazing thing: there are only two liquids, mercury and bromine. OK, gallium melts at 30°C so it's a liquid on a hot day, too. Look for elements that are typically found in the gaseous state and only the six noble gases plus five other materials are included. It is interesting that so many liquid and gaseous compounds exist when so few elements naturally occur in those states.

A substance such as water is often found in all three common states simultaneously. A glass of ice water will have solid water, liquid water and also some evaporation will form gaseous water (water vapor) that escapes from the surface of the liquid.

How is it that ice cubes can disappear from the freezer of your frost-free refrigerator? They sublime, or pass directly from solid to gas phase. It happens because the freezer removes almost all of the humidity (water vapor) from the air inside. It allows the molecules of water to leave the surface of the ice without needing to melt the ice first.

Gases

2.1 Facts about Gases

Definition: A *Gas* is a fluid comprised of a collection of atoms or molecules that are relatively far apart. To describe a gas at any point in time it is necessary to specify three quantities: volume, pressure and temperature.

Definition: A *Fluid* is a substance that can flow and will conform to the boundaries of a container in which it is confined. Fluids differ from solids in that their molecules have no long-range order.

> **More on Fluids**
>
> A fluid flows because it cannot withstand a force that is tangential to its surface. That kind of a force is called a *shearing stress*. It can be said that a fluid is a substance that flows *because* it cannot withstand a shearing stress. It can exert a force that is perpendicular to its surface, however.

All five of the gaseous elements that are not noble gases occur as diatomic molecules in nature. Two of the atoms of the gas combine, sharing an electron in a chemical bond. That is why the common way of writing the chemical symbol for these gases looks like this: H_2, N_2, O_2, F_2 and Cl_2. Of course, the noble gases are not chemically reactive so they do not readily combine with each other or any other substance. Oxygen can form the triatomic molecule O_3 (ozone) under certain conditions.

To the physicist, gases are compressible fluids. The study of fluids includes both liquids and gases but because gases are compressible, they tend to be a bit more complicated in the physical sense and have gotten quite a lot more attention than liquids.

The favorite unit of pressure used in the chip industry is the Torr, named after Galileo's personal secretary, Evangelista Torricelli, who is credited with inventing the barometer (see Chapter 4). It is the same unit as the millimeter of mercury. Torricelli found that the column of mercury (Hg) in his barometer was, on average, 760 millimeters high at sea level. Even though the metric system was invented several hundred years after Torricelli's death we must use it here because no one would recognize the measurement system used in those days.

Table A-1: Facts about Gases

- Standard Atmospheric pressure = 760 Torr = 760 mm Hg
- At 1 Torr, 1cc (cubic centimeter) of gas contains approximately 3.5×10^{16} (35 million billion) particles of gas
- Gaseous particles are in constant motion
  - Constantly colliding with each other and chamber walls
  - 400 m/sec at room temperature (typical approximate average speed)
  - Increased temperature increases velocity
- Mean Free Path: the average distance a particle travels between collisions with another particle
- Approximate mean free paths for hydrogen molecules at low pressure (in the range used in many chip processes)
  - 1 Torr   ~0.1 mm (millimeter)
  - 100 mT   ~1 mm
  - 10 mT   ~1 cm

2.2 Ions

Definition: An *Ion* is a particle that has an electrical charge.

An ion has either lost or gained one or more electrons, resulting in an excess of negative or positive electrical charge. Remember that the electron is the negative charge carrier and the proton (in the nucleus) carries the positive charge.

2.3 Plasma

Definition: *Plasma* is an ionized gas.

Many people are surprised to learn that the plasma state is the most common state of matter in the universe. The sun and stars are mostly plasma and interstellar space has lots of plasma, too. Cold Dark Matter is a new arrival on the theoretical scene, and nobody knows exactly what it is yet so we will only discuss "normal" matter here.

Plasma is used extensively in the semiconductor manufacturing process so it is important to know something about this unfamiliar state of matter.

Table A-2: Facts about Plasma

- Produced through the use of electrical energy to ionize gases
- In a plasma etcher (a semiconductor manufacturing tool),
  the gas is partially ionized
  - Typically less than 1% of the atoms and molecules are ionized
    - Sometimes only one atom in 100,000 is ionized
  - High Density plasmas such as Inductively Coupled Plasma (ICP)
    and Electron Cyclotron Resonance (ECR) plasmas may exceed
    1% ionized species
- "Glow Discharge" is another name for a plasma
- Plasmas produce highly reactive chemicals
  - Greatly reduces the amount of chemical needed
  - Facilitates some chemical reactions
  - Reduces processing time
- Ion Bombardment in Etchers
  - Provides the ability to shape the features being created
  - Energy is added to increase the rate of reaction
- Extremely clean compared to wet (liquid) processing
  - *Scrubbers* (tools that trap and neutralize process by-products)
    replace the recycling or disposal of lots of toxic liquids

2.4 Free Radicals

Definition: A *Free Radical* is a highly reactive fragment of a molecule that has one or more "free" or unpaired electrons.

A free radical is a highly reactive substance. If a CF_4 molecule is decomposed in a processing reactor, which is a common activity in the chip-making process, it might produce free radicals like CF_3, CF_2, CF.

To make matters even more complicated, the free radicals can become ionized too. In addition, free radicals can be in an excited state as described in the next section.

2.5 Excited States

Definition: Atoms or molecules are in an *Excited State* if they have extra energy added to them.

An example will help to explain excited states. Heating a material is one way of adding energy to it. In many plasma reactors the gas molecules are heated by using an electric field to accelerate stray electrons, causing them to collide with other particles.

The collisions transfer energy (heat) to the particles of gas, causing electrons to move further away from the nucleus or to break away entirely. When the electrons give up their extra energy (cool down) they radiate it in the form of light and move closer to the nucleus again.

Chemical reactions happen much more rapidly with the chemicals in an excited state. However, if a molecule is not involved in a reaction, the extra energy is usually given up rather quickly. For example, a gas may glow, emitting a characteristic color. The extra energy is being given up as photons of light (for example, a neon sign). But as long as the energy is being added to the gas, its atoms or molecules will become excited, emit light and drop to a lower energy level, become excited again, giving up the energy, over and over.

Since fast processing time is important in any manufacturing environment, the use of chemicals in an excited state is of great help in the chipmaking process.

Chemistry

3.1 Introduction

The chemistry involved in chipmaking is relatively straightforward. The number of chemicals used in the manufacturing process is fairly limited as are the types of reactions. However, a large number of elements and compounds are of importance to the discussion of the subject. So rather than try to summarize an entire chemistry course, this discussion will be confined to only the topics involved in making chips. A few examples are given for each topic to serve as a familiarization and review.

Chemical reactions are primarily needed for making the silicon wafers, depositing films on the wafers and etching the patterns into the films. Examples of deposition chemistry and plasma etch chemistry will be discussed.

The following definition is offered for the sake of completeness.

Definition: *Chemistry* is the study of matter and the many ways in which it can change.

3.2 Thin Film Deposition Chemistry

Silicon technology is the focus of this book. Let us begin with an example of a chemical reaction involving silicon.

$$Si + O_2 \rightarrow SiO_2$$

This simple chemical equation shows silicon reacting with oxygen to form silicon dioxide. The materials on the left side of the arrow are called the *reactants*; the plus sign means "reacts with." The arrow means "produces." The product(s) formed in the reaction are to the right of the arrow.

Key Point: The rate of reaction will increase if energy is added. Energy is often added by heating the reactants, but other methods are available.

This oxidation reaction will happen spontaneously in the air at atmospheric pressure and room temperature. However, the reaction goes extremely slowly under those conditions. In order to speed it up and thereby improve *throughput* (the number of wafers processed per hour), the process is done at high temperature. In the case of the growth of sacrificial oxide done early in the process, the wafer will typically be heated to more than 900°C in a furnace. Inert gas is used to flush the furnace when idle. After loading wafers inside, the chamber is evacuated prior to introducing oxygen.

For some thicker oxide growth steps used in chip fabrication, water vapor is used instead of oxygen. Surprisingly, it gives a faster oxidation rate.

Here is the reaction with water vapor:

$$Si + 2H_2O \rightarrow SiO_2 + 2H_2$$

The hydrogen comes off as a gas.

Letters in parenthesis occasionally appear in the equation to show the state of each species. The usual designations are solid (s), liquid (l) and gas (g).

Balanced Chemical Equations

In the reaction above, a "2" appears in front of the water component, and a "2" is found in front of the hydrogen term. These factors are needed to balance the chemical equation. A balanced equation has the same number of atoms of each element on both sides of the arrow.

For brevity, many chemical equations are presented in their unbalanced form.

An example of a deposited film is silicon nitride. One common chemical vapor deposition (CVD) method for depositing this film is combining ammonia gas (NH_3) with the vapor of a liquid, trichlorosilane ($SiHCl_3$).

$$NH_3 + SiHCl_3 \rightarrow Si_3N_4 + HCl$$

This reaction requires a temperature of about 700°C (above red heat). Later in the process, lower temperatures of deposition are required. Then an alternative reaction can be used, reacting silane (SiH_4) and ammonia.

$$SiH_4 + NH_3 \rightarrow Si_xN_yH_z$$

Using the energy of a plasma, this reaction can be done at around 400°C. The composition and properties of this nitride film are considerably different from the higher temperature film but are still acceptable for some applications.

Why not just react silicon and nitrogen? Because the reaction requires a temperature of 1000–1300°C to take place. That high temperature is incompatible with the process restrictions, as discussed in the text. Another small difficulty arises when nitride is needed in the process: usually all of the silicon is covered up by other films, so none is available to react with the nitrogen to form nitride. CVD to the rescue!

3.3 Plasma Etch Chemistry

Plasma processing is widely used in chipmaking. The plasma produces many highly reactive species. In essence, the production of a plasma adds energy to the gaseous chemicals through the use of an electric field. The increased rate of reaction is one of several advantages discussed in the text.

Plasma tools are used for both deposition and etch. Only the etch examples are presented here.

Etching silicon is necessary to make the features on the chip. Chlorine is often used to etch silicon.

$$Si(s) + Cl_2(g) \rightarrow SiCl_4(g)$$

The solid silicon reacts with the gaseous chlorine to produce gaseous silicon tetrachloride. The functioning of the etch reactor is described in Chapter 4.

Key Point: Plasma etch reactions must produce volatile products of reaction or the reactant chemicals will be blocked from reaching the wafer and the reaction will stop.

Often the etch reaction will include a component that will be left behind as a solid on the wafer surface. This example shows a partial sampling of how much is going on inside many plasma reactors. CF_4 (Freon 14®) and CHF_3 (Freon 113®) are the two reactants with silicon dioxide.

$$SiO_2(s) + CF_4(g) + CHF_3(g) \rightarrow SiF_4(g) + CO(g) + CO_2(g) + CF_x(s) + HF(g) + F(g) + \ ...$$

In this example, the silicon dioxide reacts with fluorine produced in the plasma, giving silicon tetrafluoride, a gas. The CF_x in the equation represents various carbon/fluorine polymers produced in the reaction. These polymers are solids and deposit on the wafer surface, forming a passivation layer that is essential to the process. The complexity of the plasma chemistry is discussed extensively in Chapters 4 and 5. This extensive and incomplete equation is shown to indicate the complexity of the reaction.

Solids

Definition: A *Solid* is a material that holds its shape and volume, and tends to resist physical deformity. The particles that make up a solid are in a fixed position within the material, often in an ordered arrangement and are relatively close together. The atoms and molecules still vibrate but do not leave their positions within the solid.

Electrical Properties of Solids

4.1 Conductors and Insulators

Definition: A *Conductor* is a material that readily conducts electricity. Electrical conductors also tend to conduct heat well.

Material Properties

It is typical to discuss the properties of a material in terms of its resistivity and the reciprocal quantity, conductivity. Resistivity is a property of a material. It is defined as the electric field intensity at a specific point in the material divided by the current density at that point. The units of resistivity are ohm-meters; however, it is far more convenient to convert to ohm-centimeters for use in the semiconductor industry and other industries. The ohm-centimeter is the resistance measured on opposite faces of a one centimeter cube of the material of interest.

Definition: An *Insulator* is a very poor conductor of electricity. Insulators are said to be highly resistive—that is, they resist the flow of current. The best insulator is a vacuum.

4.2 Semiconductors

Definition: A *Semiconductor* is a material with electrical properties midway between a conductor and an insulator. Semiconductor materials readily change electrical properties with the addition of impurities called *dopants*.

In keeping with the focus of this book, silicon will be the primary topic of discussion in this section.

Table A-3: Facts about Semiconductors

- Examples: silicon, germanium, gallium arsenide
- Respond readily to doping, the addition of impurities
- Changes in the resistivity of silicon
  - Totally pure ("intrinsic") silicon >100,000 Ohm-cm
  - 1 part per billion of arsenic ~100 Ohm-cm
  - 1 part per million of arsenic ~0.15 Ohm-cm

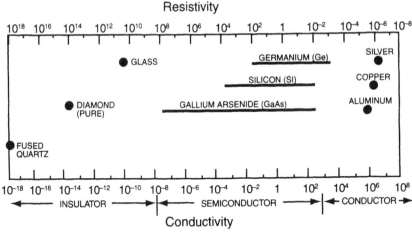

Figure A-7: Electrical Conductivities of Solids (Wolf)

Semiconductor devices are made primarily by doping the semiconducting material to adjust its electrical properties. The semiconductor may be doped with impurities that provide extra mobile negative charges (electrons) to the material, making it "n-type" or the impurities may be short of electrons and thereby provide apparent mobile positive charges (holes) to make the material "p-type." The p-type material is treated as if it had actual positive charge carriers, even though the actual physical situation is a "lack of negative charges."

Doping raises the conductivity of silicon to a level where it is considered to be an electrical conductor, (Figure A-7) so it can be used to carry current. However, its conductivity is still four orders of magnitude below that of a metal such as copper, a particularly good conductor.

4.3 pn Junctions

Definition: A *pn Junction* is single piece of semiconductor material that has been selectively doped such that adjacent regions of n-type and p-type material are formed. It is an example of a *solid state device*. The term solid state originated when transistors made from semiconducting materials replaced vacuum tubes.

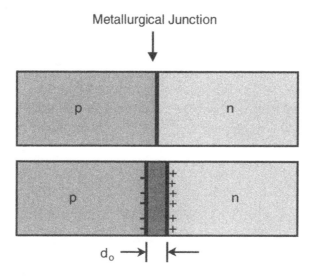

Figure A-8: pn Junction

The magic of solid state devices is revealed in the pn junction. When adjacent regions of oppositely doped silicon are formed, an interesting phenomenon occurs. At the interface, a thin area called the *depletion region* forms. The depletion region is so-called because it is depleted of charges that are free to move. In this condition it cannot conduct a current in the usual way. However, if the proper voltage is applied, as shown in Figure A-9, a current will flow quite easily.

The pn junction can serve as a *rectifier*, a device that allows current to flow only one direction through it. Figure A-9 illustrates the functioning of the junction rectifier. Note that the depletion region grows much wider with a reverse bias, resisting the flow of current through the device in that direction.

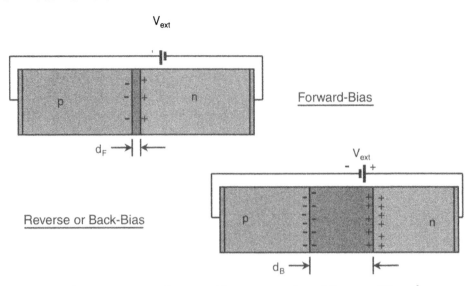

Figure A-9: Junction Rectifier, Forward and Reverse Biased

Many solid-state devices are made by combining pn junctions in creative ways. In this book, a MOSFET was made, one example of combining two pn junctions near to each other.

Electricity, Electric and Magnetic Fields

5.1 Electric Charges and Fields

Definition*: Electric Charge* is an intrinsic property of matter. Since matter is made up of atoms that contain electrons (negative) and protons (positive), the charges are always there. Many natural and artificial conditions exist which allow us to work with charges (usually electrons) and to take advantage of the energy available.

Definition*:* An *Electric Field* exists in the region around a charged particle or a group of charges. It describes the amount of force exerted on other charges that enter the region. The field is invisible to everything except another charged particle.

Electric Field Strength

The electric field strength is the force exerted on a test charge at a given point within the field. The field strength decreases rather rapidly with distance, as the test charge is moved away from the source of the field; this is an example of the *inverse square law* where the field strength falls off as the inverse square of the distance from the source increases.

A different situation exists in a *parallel-plate capacitor*. The electric field is uniform between the capacitor plates (except at the edges, but that can be conveniently ignored). The parallel-plate capacitor is a familiar example used throughout the text.

Field Theory

Field theory explains the "action at a distance" effect experienced by charged objects; they exert a force on each other (like charges repel and unlike charges attract) from relatively far away. Before Field Theory was formulated many scientists thought that an unseen fluid called *ether* existed that transmitted the electrical force from one object to the other. Late in the Nineteenth Century, Michelson and Morley proved that there is no ether and paved the way for electric field theory.

5.2 Electric Current

Definition: An *Electric Current* is the flow of electric charges. A net flow in one direction of any charged particles could be called an *electric current*, but for practical reasons, this discussion will be confined to the flow of electrons, usually within a conducting medium.

One confusing issue that should be mentioned, is that the convention adopted for current flow in an electric circuit is from the positive pole of the source, such as a battery, to the negative pole. The actual movement of the electrons goes in the opposite direction. However, everything still works, including the math, if the pretence of flowing positive charges is maintained.

Do not confuse conventional electric current with, say, the movement of positive ions within a plasma, a real example of the movement of positive charges. The physics is usually quite different from that needed for this discussion (and a lot harder).

Current flow is often compared to the flow of water through a pipe and the analogy is quite accurate in many ways. For example, electric current is analogous to the volume of water flowing through a pipe. Conducting wires or other materials are analogous to the pipe. If the pipe is too small for the amount of water being fed into it, then it slows the flow just as a fine wire will resist a high electrical current flow; a "resistor" is often made of fine wire.

Definition: A *Potential Difference* or a *Voltage* must exist in order for electrons to flow. The potential difference is the equivalent of pressure in a pipe that pushes water through it. The water will not flow without pressure to push it and neither will electrons flow without a potential difference.

5.3 Magnetic Field

Definition: A *Magnetic Field* is similar to an electric field; it is an intrinsic property of charged particles. However, it also exists in the region around a permanent magnet (a material with its electrons lined up in fixed positions) or in a conductor with current flowing through it.

Electromagnets, such as metal bars with current carrying wire wrapped around them, act just like permanent bar magnets.

Magnetic fields interact with charged particles somewhat differently than electric fields. These differences will be pointed out in the discussions of the tools that use magnetic fields.

Plasma Etch Supplement to Chapter 4

Plasma Etcher Theory

Plasma is an ionized gas. How does an etch reactor make a gas into a plasma?

Free electrons in a plasma reactor are accelerated by an electric field; they collide with electrons orbiting the nuclei of atoms and molecules, knocking some of them off and producing the cascade effect. Knocking off an electron will produce a positive ion or one of the other components of the plasma discussed below. In other cases, an atom or molecule of gas will trap a free electron, producing a negative ion. Figure B-1 attempts to show the complicated "chemical soup" that exists in a plasma reactor.

Of course, electrons collide with everything in the plasma reactor including the reactor walls. Those collisions do not contribute much of anything to making the plasma, so they will not be discussed here. Several interesting references may be found in the bibliography that fully discuss plasma physics and chemistry.

Key Point: Plasma is produced entirely by electron collisions.

That key point is a bit surprising when one considers that an electron is more than 1,830 times lighter than a proton or neutron, the components of the nucleus (see Appendix A). Many of the atoms in the plasma are fairly massive, with numerous protons and neutrons in the nucleus. Why don't the nuclei dominate the collisions rather than electrons? There are a couple of reasons.

The first reason is just a part of the amazing nature of atoms themselves. It has been said that if an atom were the size of a cathedral the nucleus would be no bigger than a fly—but a fly that weighs thousands of times more than the cathedral. All of that extra space is the domain of the electrons belonging to that atom. The chances are much greater for electron collisions than for an electron-nucleus collision.

Second, the electrons are extremely light compared to the ions because the nuclei of the ions are so massive. Plasma reactors are designed to take advantage of that fact and really put the power into whipping the electrons around. Accelerating the ions is important, too, for some applications, but the purpose in that case is to bombard the wafer, not to make plasma.

Another interesting thing happens in a plasma. The electron collisions produce ions but also break up molecules into their component parts of atoms, radicals, excited states, and so on. The gas molecules are bound together by covalent bonds, which consist of two shared electrons for each bond (there are double and triple bonds for some molecules). If one or more of those electrons is knocked away, then sometimes, rather than producing an ion, the bond will be broken and the molecule comes apart. This action produces the "chemical soup" that exists inside the etch reactor. Figure B-1 illustrates some of the chemical composition of the plasma.

$$SiO_2 + CF_4 + CH_2F_2 \rightarrow SiF_4 + CO$$

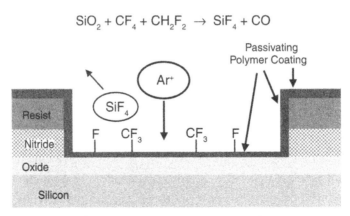

Figure B-1: Chemical Soup in a Plasma

Chemical Reactions

The ease with which a chemical reaction takes place depends a lot upon the strength of the bonds holding the reactants together. For example, the strength of the chemical bonds holding the silicon to the oxygen in silicon dioxide is more than double the strength of the bond holding the silicon atoms to each other in single crystal silicon. Needless to say, silicon dioxide etches much more slowly than silicon in a given set of conditions. In this section, the methods used to increase the etch rate of silicon dioxide are discussed.

Plasma Etch Process Requirements

Several types of etch reactors are used in manufacturing today (see Chapter 4). Etch technologies are referred to by the reaction chamber designs they use (parallel plate, inductively coupled, ECR, etc.). However, no matter what approach is used, the same result must be produced on the wafer. Here are some of the plasma etch process requirements.

Here are some of the plasma etch process requirements:

- High Etch Rate for High Throughput: Time is money on the manufacturing line. The rate at which wafers move through the line must be maximized.

- Uniformity: A highly uniform rate of film removal across the wafer's surface is important so that the etch is completed at all points on the wafer at almost the same instant. There is usually some degree of attack of the underlying film; if the film of interest is cleared faster at one region on the wafer, then more of the underlying film will be removed from that area while the etch is completed on the rest of the wafer. There is a strong possibility of damaging the devices in that region because of over-etching.

- Selectivity: The objective is to etch the film of interest and not attack any other films on the wafer. Unfortunately, that is rarely possible. Strategies that will minimize the attack of the underlying film and the masking material will be discussed for each etch.

- Sidewall Profile Control: The ability to control the shape of the features being etched is critical for most etches. The sidewall slope and other key shapes will be discussed for each application.

There are many additional requirements specific to particular applications. These are usually the highest priority. The references in the Bibliography offer ideas for further reading.

Bibliography

1. Cropper, William H., *Great Physicists: The Life and Times of Leading Physicists from Galileo to Hawking*, New York, Oxford University Press, 2001.

2. Sze, S.M., *Physics of Semiconductor Devices*, Second Edition, John Wiley & Sons, 1981.

3. Halliday, David, Robert Resnick, and Jearl Walker, *Fundamentals of Physics, Extended*, Fifth Edition, John Wiley & Sons, 1997.

4. Brown, Theodore L., and H. Eugene LeMay, Jr., *Chemistry the Central Science*, Prentice-Hall, Inc., 1977.

5. *The Oxford American Dictionary and Language Guide*, Oxford University Press, 1999.

6. Wolf, S., *Microchip Manufacturing*, Lattice Press, 2004.

7. Sze, S.M., *VLSI Technology*, Bell Telephone Laboratories, Incorporated, 1983.

8. Ghandhi, Sorab K., *VLSI Fabrication Principles, Silicon and Gallium Arsenide*, John Wiley & sons, Inc., 1983.

9. *Semiconductor & Integrated Circuit Fabrication Techniques*, Fairchild Corporation, 1979.

10. Manos, Dennis M., and Daniel L. Flamm, *Plasma Etching: An Introduction*, Academic Press, Inc., 1989.

11. Lieberman, Michael A., and Allan J. Lichtenberg, *Principles of Plasma Discharges and Materials Processing*, John Wiley & Sons, Inc., 1994.

12. Chapman, Brian, *Glow Discharge Processes, Sputtering and Plasma Etching*, John Wiley & Sons, Inc., 1980.

13. Rossnagel, Stephen M., Jerome J. Cuomo, and William D. Westwood, *Handbook of Plasma Processing Technology, Fundamentals, Etching, Deposition, and Surface Interactions*, Noyes Publications, 1990.

14. Heynes, Michael, and Anne K. Miller, *Integrated Circuit Synopsis*, Third Edition, Semiconductor Services, 2002.

15. Heynes, Michael, and Anne K. Miller, *Semiconductor Terminology – Graphic Glossary of Terms*, Fourth Edition, Semiconductor Services, 2002.

16. *Making the Microchip: At the Limits III Video Training Course*, Semiconductor Services, 2004.

17. Evans, K., A. Miller and P. Van Zant, *Cleanroom Technology Manual*, Second Edition, Semiconductor Services. 1990.

18. Van Zant, Peter, *Microchip Fabrication*, Fifth Edition, McGraw-Hill, 2004.

19. Gise, Peter E., Blanchard, Richard, *Semiconductor and Integrated Circuit Fabrication Techniques*, Reston Publishing Company, Inc., 1979.

20. Van Nostrand's Scientific Encyclopedia, Fifth Ed., Litton Educational Publishing, Inc., 1976.

21. CRD Handbook of Chemistry and Physics, 76th Ed., 1995–1996, CRC Press, Inc., 1995.

22. Plummer, James D., Deal, Michael D., Griffin, Peter B., *Silicon VLSI Technology*, Prentice Hall, Inc., 2000.

23. *The Semiconductor Picture Dictionary*, DM Data, Inc., 1989.

Index

TERM

This Agreement will remain in effect until terminated pursuant to the terms of this Agreement. You may terminate this Agreement at any time by removing from Your system and destroying the CD-ROM Product. Unauthorized copying of the CD-ROM Product, including without limitation, the Proprietary Material and documentation, or otherwise failing to comply with the terms and conditions of this Agreement shall result in automatic termination of this license and will make available to Elsevier Science legal remedies. Upon termination of this Agreement, the license granted herein will terminate and You must immediately destroy the CD-ROM Product and accompanying documentation. All provisions relating to proprietary rights shall survive termination of this Agreement.

LIMITED WARRANTY AND LIMITATION OF LIABILITY

NEITHER ELSEVIER SCIENCE NOR ITS LICENSORS REPRESENT OR WARRANT THAT THE INFORMATION CONTAINED IN THE PROPRIETARY MATERIALS IS COMPLETE OR FREE FROM ERROR, AND NEITHER ASSUMES, AND BOTH EXPRESSLY DISCLAIM, ANY LIABILITY TO ANY PERSON FOR ANY LOSS OR DAMAGE CAUSED BY ERRORS OR OMISSIONS IN THE PROPRIETARY MATERIAL, WHETHER SUCH ERRORS OR OMISSIONS RESULT FROM NEGLIGENCE, ACCIDENT, OR ANY OTHER CAUSE. IN ADDITION, NEITHER ELSEVIER SCIENCE NOR ITS LICENSORS MAKE ANY REPRESENTATIONS OR WARRANTIES, EITHER EXPRESS OR IMPLIED, REGARDING THE PERFORMANCE OF YOUR NETWORK OR COMPUTER SYSTEM WHEN USED IN CONJUNCTION WITH THE CD-ROM PRODUCT.

If this CD-ROM Product is defective, Elsevier Science will replace it at no charge if the defective CD-ROM Product is returned to Elsevier Science within sixty (60) days (or the greatest period allowable by applicable law) from the date of shipment.

Elsevier Science warrants that the software embodied in this CD-ROM Product will perform in substantial compliance with the documentation supplied in this CD-ROM Product. If You report significant defect in performance in writing to Elsevier Science, and Elsevier Science is not able to correct same within sixty (60) days after its receipt of Your notification, You may return this CD-ROM Product, including all copies and documentation, to Elsevier Science and Elsevier Science will refund Your money.

YOU UNDERSTAND THAT, EXCEPT FOR THE 60-DAY LIMITED WARRANTY RECITED ABOVE, ELSEVIER SCIENCE, ITS AFFILIATES, LICENSORS, SUPPLIERS AND AGENTS, MAKE NO WARRANTIES, EXPRESSED OR IMPLIED, WITH RESPECT TO THE CD-ROM PRODUCT, INCLUDING, WITHOUT LIMITATION THE PROPRIETARY MATERIAL, AN SPECIFICALLY DISCLAIM ANY WARRANTY OF MERCHANTABILITY OR FITNESS FOR A PARTICULAR PURPOSE.

If the information provided on this CD-ROM contains medical or health sciences information, it is intended for professional use within the medical field. Information about medical treatment or drug dosages is intended strictly for professional use, and because of rapid advances in the medical sciences, independent verification of diagnosis and drug dosages should be made.

IN NO EVENT WILL ELSEVIER SCIENCE, ITS AFFILIATES, LICENSORS, SUPPLIERS OR AGENTS, BE LIABLE TO YOU FOR ANY DAMAGES, INCLUDING, WITHOUT LIMITATION, ANY LOST PROFITS, LOST SAVINGS OR OTHER INCIDENTAL OR CONSEQUENTIAL DAMAGES, ARISING OUT OF YOUR USE OR INABILITY TO USE THE CD-ROM PRODUCT REGARDLESS OF WHETHER SUCH DAMAGES ARE FORESEEABLE OR WHETHER SUCH DAMAGES ARE DEEMED TO RESULT FROM THE FAILURE OR INADEQUACY OF ANY EXCLUSIVE OR OTHER REMEDY.

U.S. GOVERNMENT RESTRICTED RIGHTS

The CD-ROM Product and documentation are provided with restricted rights. Use, duplication or disclosure by the U.S. Government is subject to restrictions as set forth in subparagraphs (a) through (d) of the Commercial Computer Restricted Rights clause at FAR 52.22719 or in subparagraph (c)(1)(ii) of the Rights in Technical Data and Computer Software clause at DFARS 252.2277013, or at 252.2117015, as applicable. Contractor/Manufacturer is Elsevier Science Inc., 655 Avenue of the Americas, New York, NY 10010-5107 USA.

GOVERNING LAW

This Agreement shall be governed by the laws of the State of New York, USA. In any dispute arising out of this Agreement, you and Elsevier Science each consent to the exclusive personal jurisdiction and venue in the state and federal courts within New York County, New York, USA.